HISTORICALLY BLACK COLLEGES AND UNIVERSITIES

MARYBETH GASMAN'S
PREVIOUS PUBLICATIONS

Envisioning Black Colleges: A History of the United Negro College Fund (2007).

Fund-Raising from Black College Alumni: Successful Strategies for Supporting Alma Mater (2003), with coauthor Sibby Anderson-Thompkins. Winner of the H. S. Warwick Award for Outstanding Published Scholarship on Philanthropy.

Understanding Minority Serving Institutions (2008), with coeditors Benjamin Baez and Caroline Sotello Turner.

Uplifting a People. Essays on African American Philanthropy and Education (2005), with coeditor Kate Sedgwick. Winner of the Association for Fundraising Professional's Skytone Ryan Research Prize.

Charles S. Johnson: Leadership behind the Veil in the Age of Jim Crow (2003),with coauthor Patrick Gilpin.

Gender and Educational Philanthropy: New Perspectives on Funding, Collaboration, and Assessment (2007), with coeditor Alice C. Ginsberg.

Philanthropy, Volunteerism, and Fundraising (2007), with coeditor Andrea Walton.

HISTORICALLY BLACK COLLEGES AND UNIVERSITIES

TRIUMPHS, TROUBLES, AND TABOOS

Edited by

Marybeth Gasman and Christopher L. Tudico

palgrave
macmillan

HISTORICALLY BLACK COLLEGES AND UNIVERSITIES

First published in 2008 by PALGRAVE MACMILLAN® in the United States – a division of St. Martin's Press LLC, 175 Fifth Avenue, New York, NY 10010.

Where this book is distributed in the UK, Europe and the rest of the world, this is by Palgrave Macmillan, a division of Macmillan Publishers Limited, registered in England, company number 785998, of Houndmills, Basingstoke, Hampshire RG21 6XS.

Palgrave Macmillan is the global academic imprint of the above companies and has companies and representatives throughout the world.

Palgrave® and Macmillan® are registered trademarks in the United States, the United Kingdom, Europe and other countries.

ISBN-13: 978-0-230-60273-1
ISBN-10: 0-230-60273-8

Library of Congress Cataloging-in-Publication Data

Historically black colleges and universities : triumphs, troubles, and taboos / edited by Marybeth Gasman & Christopher L. Tudico.
    p.   cm.
ISBN 0-230-60273-8
1. African American universities and colleges.   2. African Americans—Education (Higher)
I. Gasman, Marybeth.   II. Tudico, Christopher L.

LC2781.H573   2009
378.73089'96073—dc22   2008021759

A catalogue record of the book is available from the British Library.

Design by Macmillan Publishing Solutions

First edition: December 2008

10 9 8 7 6 5 4 3 2 1

Printed in the United States of America.

CONTENTS

TABLES AND ILLUSTRATIONS

ACKNOWLEDGMENTS

In the fall of 2005, I taught a seminar at the University of Pennsylvania's Graduate School of Education on historically Black colleges and universities (HBCUs) in which Christopher Tudico was a student. As part of the class, students were given an opportunity to write a chapter for this edited book. My goal was to teach them about the writing, research, and the publishing process. Seven of the 13 chapters in this book began in this seminar on Black colleges. I am very proud of these students and how far their work has come over the course of two years. The balance of the chapters was written by colleagues throughout the country who have an interest in HBCUs; many of these authors are young scholars, at the beginning of their careers themselves.

Once the book was under way, I brought Christopher on as a coeditor; he had an interest in HBCUs and serves as my research assistant. I thought that editing a book would be a good skill for him to learn as he works to become a university professor. Christopher has been a wonderful coeditor, spending many, many hours reading these chapters and looking for ways to make them better and more useful to the reader— be they a scholar, practitioner, or policymaker.

When writing a book there are many people to thank. We are grateful to our faculty and doctoral student colleagues at the University of Pennsylvania, who contribute to a wonderfully rich intellectual environment that enables all to thrive! We are thankful to the individual authors who completed many sets of revisions and answered our questions and concerns throughout the writing process. Our editors at Palgrave, Amanda Moon and Julia Cohen, were helpful in shaping the idea for this book as were the anonymous reviewers of the manuscript. We are also thankful to the faculty members, staff, and students at HBCUs, as these institutions, despite the challenges some of them face, contribute greatly to higher education for African Americans (and many others) in the United States. The value-added impact of an education at an HBCU is quite remarkable!

On a personal note, I am thankful to my husband, Edward, and daughter Chloe for their continual support of my research. Christopher wishes to thank his parents, Louis and Emma Tudico, and his sister Angela.

Marybeth Gasman and Christopher L. Tudico
Philadelphia, Pennsylvania

Introduction

Marybeth Gasman and Christopher L. Tudico

Historically Black colleges and universities (HBCUs) play a vital role in the education of African Americans in the United States. For nearly 150 years, these institutions have trained the leadership of the Black community, graduating the nation's African American teachers, doctors, lawyers, scientists, and college faculty. Regardless, through most of their history, Black colleges have faced great scrutiny; their very existence has been questioned. Early on, those who fundamentally believed that African Americans were not worthy of education challenged Black colleges (or tried to dilute their curricula); since *Brown v. Board*, those who believe that these institutions perpetuate segregation have called for their closure or at the very least, their merger with nearby historically White institutions (HWIs). Despite the controversy they face, Black colleges continue their commitment to racial uplift and education.

Historical Background

A few Black colleges appeared immediately before the Civil War, such as Lincoln and Cheyney universities in Pennsylvania and Wilberforce in Ohio (Butchart, 1980). With the end of the Civil War, the daunting task of providing education to over four million formerly enslaved people was shouldered by both the federal government, through the Freedmen's Bureau, and many northern church missionaries. As early as 1865, the Freedmen's Bureau began establishing Black colleges, resulting in staff and teachers with primarily military backgrounds. During the postbellum period, most Black colleges were so in name only; these institutions generally provided primary and secondary education, a feature that was true of most historically White colleges during the first decades of their existence.

As noted, religious missionary organizations—some affiliated with northern White denominations such as the Baptists and Congregationalists and some with Black churches such as the African Methodist Episcopal and the African Methodist Episcopal Zion—were actively working with the Freedmen's Bureau. Two of the most prominent White organizations were the American Baptist Home Mission Society and the American Missionary Association, but there were many others as well. These White northern missionary societies founded Black colleges such as Fisk University in Nashville, Tennessee, and Spelman College in Atlanta, Georgia. The benevolence of the missionaries was tinged with self-interest and sometimes racism. Their goals in establishing these colleges were to Christianize the freedmen (i.e., convert former-enslaved people to their brand of Christianity) and to rid the country of the "menace" of uneducated African Americans (Anderson, 1988). Among the colleges founded by Black denominations were Morris Brown in Georgia, Paul Quinn in Texas, and Allen University in South Carolina. Unique among American colleges, these institutions were founded by African Americans for African Americans (Anderson, 1988). Because these institutions relied on less support from Whites, they were able to design their own curricula; however, they were also more vulnerable to economic instability.

With the passage of the second Morrill Act in 1890, the federal government again took an interest in Black education, establishing public Black colleges. This act stipulated that those states practicing segregation in their public colleges and universities would forfeit federal funding unless they established agricultural and mechanical institutions for the Black population. Despite the wording of the Morrill Act, which called for the equitable division of federal funds, these newly founded institutions received less funding than their White counterparts and thus had infe-rior facilities. Among the 17 new "land grant" colleges were institutions such as Florida Agricultural and Technical University and Alabama Agricultural and Mechanical University.

At the end of the nineteenth century, private Black colleges had exhausted funding from missionary sources. Simultaneously, a new form of support emerged, that of White northern industrial philanthropy. Among the leaders of industry who initiated this type of support were John D. Rockefeller, Andrew Carnegie, Julius Rosenwald, and John Slater. These industry captains were motivated by both Christian benevo-lence and a desire to control all forms of industry (Anderson, 1988; Watkins, 2001). The organization making the largest contribution to Black education was the General Education Board (GEB), a conglomeration of northern White philanthropists, estab-lished by John D. Rockefeller Sr. but spearheaded by John D. Rockefeller Jr. Between 1903 and 1964, the GEB gave over $63,000,000 to Black colleges, an impressive fig-ure, but nonetheless only a fraction of what they gave to White institutions (Gasman, 2007). Regardless of their personal motivations, the funding system that these indus-trial moguls created showed a strong tendency to control Black education for their benefit, to produce graduates who were skilled in the trades that served their own enterprises (commonly known as industrial education) (Anderson, 1988; Watkins, 2001). Above all, the educational institutions they supported were extremely careful not to upset the segregationist power structure that ruled the South by the 1890s. Black colleges such as Tuskegee and Hampton were showcases of industrial education (Lewis, 1994). At these colleges, students learned how to shoe horses, make dresses, cook, and clean under the leadership of individuals like Samuel Chapman Armstrong (Hampton) and Booker T. Washington (Tuskegee). The philanthropists' support of industrial education was in direct conflict with many Black intellectuals who favored a liberal arts curriculum. Institutions such as Fisk, Dillard, Howard, Spelman, and Morehouse were more focused on the liberal arts curriculum favored by W. E. B. Du Bois than on Booker T. Washington's emphasis on advancement through labor and self-sufficiency (Lewis, 1994). Whatever the philosophical disagreements may have been between Washington and Du Bois, the two educational giants did share a goal of educating African Americans and uplifting their race. Their differing approaches might be summarized as follows: Washington favored educating Blacks in the industrial arts so they might become self-sufficient as individuals, whereas Du Bois wanted to create an intellectual elite in the top ten percent of the Black population (the "talented tenth") to lead the race as a whole toward self-determination (Lewis, 1994).

Beginning around 1915, there was a shift in the attitude of the industrial philanthropists, who started to turn their attention to those Black colleges that emphasized the liberal arts. Realizing that industrial education could exist side by side with a more academic curriculum, the philanthropists opted to spread their money (and therefore their influence) throughout the educational system (Anderson, 1988). The pervasive influence of industrial philanthropy in the early twentieth century created a conservative environment on many Black college campuses—one that would seemingly tolerate only those administrators (typically White men) who accommodated segregation. But attention from the industrial philanthropists was not necessarily welcomed by institutions like Fisk University, where rebellions ensued against autocratic presidents who were assumed by students to be puppets of the philanthropists (Anderson, 1988). In spite of these conflicts, industrial philanthropists provided major support for private Black colleges up until the late 1930s.

At this time, the industrial philanthropists turned their attention elsewhere. In response, Frederick D. Patterson, then president of the Tuskegee Institute, suggested that the nation's private Black colleges join together in their fund-raising efforts. As a result, in 1944, the presidents of 29 Black colleges created the United Negro College Fund (UNCF). The UNCF began solely as a fund raising organization but eventually took on an advocacy role as well (Gasman, 2007).

Until the *Brown v. Board of Education* decision in 1954, both public and private Black colleges in the South remained segregated by law and were the only educational option for African Americans. Although most colleges and universities did not experience the same violent fallout from the *Brown* decision as southern public primary and secondary schools, they were greatly affected by the decision. The Supreme Court's landmark ruling meant that Black colleges would be placed in competition with White institutions in their efforts to recruit Black students. With the triumph of the idea of integration, many questioned whether Black colleges should exist, and labeled them vestiges of segregation (Gasman, 2007). However, desegregation proved slow, with public Black colleges maintaining their racial makeup well into the current day. In the state of Mississippi, for example, the *Fordice* case was mired in the court system for almost 25 years, with a final decision rendered in 2004. The case, which reached the United States Supreme Court, asked whether Mississippi had met its affirmative duty under the Fourteenth Amendment's Equal Protection Clause to dismantle its prior dual university system. Despite ample evidence to the contrary, the high court decided that the answer was yes. Although the *Fordice* case only applied to those public institutions within the 5th District, it had a rippling effect within most southern states, resulting in stagnant funding levels for public Black colleges and limited inroads by African Americans into HWIs (Brown, 1999).

After the *Brown* decision, private Black colleges, which have always been willing to accept students from all backgrounds if the law would allow, struggled to defend issues of quality in an atmosphere that labeled anything all-Black inferior. Many Black colleges also suffered from "brain drain" as HWIs in the North and some in the South made efforts to attract the top ten percent of their students to their institutions once racial diversity became valued within higher education.

The Black college of the 1960s was a much different place than that of the 1920s. The leadership of the institutions switched from White to Black, and because Blacks had more control over funding, there was greater tolerance for dissent and Black self-determination. On many public and private Black college campuses throughout the South, students staged sit-ins and protested against segregation and its manifestations throughout the region. Most prominent were the four Black college students from North Carolina A&T who refused to leave a segregated Woolworth lunch counter in 1960.

During this decade, once again, the federal government took a greater interest in Black colleges. In an attempt to provide clarity, the 1965 Higher Education Act defined a Black college as "any . . . college or university that was established prior to 1964, whose principal mission was, and is, the education of black Americans." The recognition of the uniqueness of Black colleges implied in this definition has led to increased federal funding for these institutions over the last 40-plus years.

Another federal intervention on behalf of Black colleges took place in 1980 when President Jimmy Carter signed Executive Order 12232, which established a national program to alleviate the effects of discriminatory treatment of African Americans and to strengthen and expand Black colleges to provide quality education. Since this time, every U.S. president has provided funding to Black colleges through this program. President George H. W. Bush followed up on Carter's initiative in 1989, signing Executive Order 12677, which created the Presidential Advisory Board on HBCUs to advise the president and the secretary of education on the future of these institutions.

Currently, over 300,000 students attend the nation's 105 historically Black colleges (40 public four-year, 11 public two-year, 49 private four-year, and 5 private 2-year institutions). This amounts to 16% of all African American college students. Overall, the parents of Black students at Black colleges have much lower incomes than those of parents of Black students at HWIs (Wenglinsky, 1999). However, many researchers who study Black colleges have found that African Americans who attend Black colleges have higher levels of self-esteem and find their educational experience more nurturing (Brown and Freeman, 2004). Moreover, graduates of Black colleges are more likely to continue their education and pursue graduate degrees than their counterparts at HWIs (Wenglinsky, 1999). Despite the fact that only 16% of African American college students attend Black colleges, these institutions produce the majority of America's African American professional leaders (AAUP report, 1995).

Black colleges in the twenty-first century are remarkably diverse and serve a wide range of populations. Although most of these institutions maintain their historically Black traditions, on average 13% of their students are White. Because of their common mission (that of racial uplift), they are often lumped together and treated as a monolithic entity, causing them to be unfairly judged by researchers, the media, and policy makers. Just as HWIs are varied in their mission and quality, so too are the nation's Black colleges. Today, the leading Black colleges cater to those students who could excel at any top tier institution regardless of racial makeup. Other institutions operate with the needs of Black students in the surrounding region in mind. And some maintain an open enrollment policy, reaching out to those students who would have few options elsewhere in the higher education system.

UNDERSTANDING CONTEMPORARY BLACK COLLEGES

In an effort to explain the continued existence of Black colleges and understand their importance in contemporary society, numerous scholars have contributed to an expanding body of knowledge on the institutions. This scholarship has grown in quantity in the past five years. A few of the studies that have been produced fall into the category of institutional cataloging: encyclopedic lists of the various colleges with little context or interpretation (Cohen, 2000; Jackson, 2003). A more interesting category is the subject-specific work. The majority of new scholarship falls into this category, which features articles on law, including the well-known desegregation efforts in southern states (Brown, 2001; Brown et al., 2001; Brown and Freeman, 2004). Scholars have also conducted research on faculty issues, such as preparation, governance, and job satisfaction (Foster, 2001; Johnson, 2001; Minor, 2005; Perna, 2001). The largest share of the new scholarship on Black colleges pertains to students, specifically graduates' earning potential, social and cultural capital acquired as a result of attending a Black college, student success, college choice, and identity formation (Bennett and Xie, 2003; Bridges et al., 2005; Brown and Davis, 2001; Fleming, 2001; Freeman, 2002; Freeman and Cohen, 2001; Hall, 2005; Harper, 2004; Merisotis and McCarthy, 2005; Sissoko, 2005; Tabbye et al., 2004).

Despite the wealth of new research on Black colleges, there are topics that remain untouched and accomplishments that go unnoticed by the scholarly community. Some of these topics have remained off-limits because of their sensitivity and potential for controversy within the African American and majority communities, others have been held captive by a lack of data, and still others address issues that are at the margins among the higher education community. The chapters in this edited volume focus on topics that deserve further attention and that will push students, scholars, policymakers, and Black college administrators to reexamine their perspectives on and perceptions of Black colleges. The essays range from an exploration of the debilitating effects of colorism and classism on Black college campuses to an examination of the troubling sexual habits of some Black college undergraduate students to a long overdue historical investigation of McCarthyism within the Black college environment. In addition to these more taboo topics, the book addresses many of the successes, or triumphs of these historic institutions. These essays include an examination of successful retention strategies among highly transient student populations, an exploration of the role of Black colleges in producing the largest share of the nation's African American female scientists, and a study of the impact that Black colleges have on preparing African Americans for graduate school within Ivy League institutions. In exploring these issues, the essays draw upon a broad spectrum of research methods, including case study, historical research, legal analysis, and quantitative and qualitative approaches.

This edited volume is divided into three sections as indicated by the title: *Triumphs, Troubles,* and *Taboos.* The first section, *Triumphs,* focuses on examples of success within the Black college environment. However, these successes are nuanced and thus the authors present the complexities pertaining to their subject matter. Chapter 1, titled "For *Alma Mater* and the Fund: The United Negro College Fund's National Pre-Alumni Council and the Creation of the Next Generation of Donors," gets at the heart of a major problem within the Black college context—that of alumni giving. Black colleges lag behind their historically White counterparts in the area of alumni for two reasons: first, African Americans have just recently reached similar income levels to White Americans; and second, over the course of their existence, Black colleges have relied on corporate and foundation contributions rather than asking their alumni to support their alma maters (Gasman and Anderson-Thompkins, 2003). In this chapter, Noah Drezner investigates the work of the UNCF's National Pre-Alumni Council (NPAC). Since its creation in 1958, NPAC and its campus affiliates have worked with students to instill an ethic of philanthropy in them from the minute they step foot on their Black college campus. Drezner shows us that active participation in NPAC leads to more active alumni participation in fund-raising, which is vital to the future success and stability of the nation's Black colleges.

In "On Firm Foundations: African American Black College Graduates and their Doctoral Student Development in the Ivy League," Pamela Felder Thompson steers clear of the traditional deficit-laden conversations about African Americans and explores student success. In particular, she focuses on the Black college undergraduate experiences of African Americans who are now pursuing doctoral degrees at Ivy League institutions. Through the experiences of four graduates, Felder Thompson examines the way that Black colleges prepare their students for graduate school. Of course, there has been ample empirical evidence on the nurturing and supportive environments at Black colleges (Fleming, 2001; Harper, 2004); however, this chapter includes the personal stories and voices of students—something often missing from larger quantitative studies. Felder Thompson also challenges elite HWIs to develop alliances with Black colleges in an effort to increase the presence of African American students.

In Chapter 3, Shannon Gary discusses Bennett and Spelman colleges' role in producing African American females who pursue PhDs in the sciences. Much like Felder Thompson's research, the chapter explores the undergraduate education at Black colleges and, in particular, student success in the sciences. Gary's work is significant in that in 2001, African American women received 66% of the bachelor's degrees awarded to African Americans in all fields, but only 43% in earth, atmospheric, and ocean sciences and 58% in physical sciences (NSF, 2004). Only in biological sciences are African American women relatively overrepresented, as African American women received 72% of bachelor's degrees awarded to African Americans in 2001 (NSF, 2004). Of importance, according to Perna et al. (2007), analyses of Integrated Postsecondary Education Data System (IPEDS) data reveal that 33% of the bachelor's degrees awarded to Black women in Science, Technology, Engineering, and Math (STEM) fields in 2004 were from HBCUs. Gary explores how Bennett and Spelman colleges produce a disproportionate number of the African American women who pursue degrees in the graduate sciences, focusing specifically on issues of self-image, self-confidence, and self-efficacy.

Gaetane Jean-Marie, in her chapter, "Social Justice, Visionary and Career Project: The Discourses of Black Women Leaders at Black Colleges," examines the lives of African American female leaders at Black colleges. She is interested in both the personal and professional lives of these women, focusing on how they connect their experiences as educational leaders with the development of social and educational capital within Black colleges and the Black community as a whole. One of the most significant aspects of this chapter is its focus on female leaders at Black colleges. By and large, the Black college literature tends to focus on male leadership, ignoring gender and gender relations (Gasman, 2007). Of course, some of this emphasis is due to the lack of female leadership of these institutions in the past years; however, in recent years, we have seen an influx of women in key administrative positions. Jean-Marie sheds light on the perspectives of these new leaders within the context of the Black college, focusing on their visionary leadership and commitment to social justice.

The second section of the book pertains to the *Troubles* that Black colleges have faced from their inception through the current day. And, as with the successes discussed in the previous section, these troubles are multidimensional and multifaceted. Patricia C. Williams's chapter "McCarthyism's Effect on Black Colleges in Pennsylvania: A Historical Case Study of Cheyney and Lincoln Universities," leads this section. Williams looks at a well-researched topic—that of McCarthyism—but she does so within the Black college environment. Pressures on presidents to fire faculty members accused of Communist behavior were common during the 1950s; however, these pressures manifested in a much different way within the Black college setting. Presidents were often faced with supporting civil rights over civil liberties. In addition, these struggling institutions did not have large endowments to fall back on when philanthropists threatened to pull their money if Communist-affiliated faculty and staff were not let go.

In Chapter 6, titled "The Forgotten GI: The Servicemen's Readjustment Act and Black Colleges, 1944–1954," Meghan Wilson examines an educational policy that has received significant coverage by historians seeking to understand its impact on historically White colleges. However, as Wilson notes, the GI Bill, as it has come to be known, has received little to no attention when it comes to its impact on the nation's Black colleges. In the chapter herein, she definitively documents the increased enrollment as a result of the GI Bill. Already financially strapped because of years of underfunding within a system of "separate but equal," Black colleges faced myriad physical plant issues because of an influx of veterans after Congress passed the GI Bill. Wilson provides a starting point for future research on the GI Bill and its impact on African Americans.

In Chapter 7, Mark Giles uses Critical Race Theory to examine the shootings that took place at Jackson State University on May 14, 1970. This event, which included the killing of two unarmed African American men, has been overshadowed in historical literature by the Kent State shootings, which happened ten days earlier. Although the "details" of the Jackson State shootings have been captured in several publications, Giles contends that this tragedy cannot be examined in a vacuum and as such contextualizes the event, considering the racial, social, and political atmosphere of the United States during the 1970s as well as Southern culture. Of particular interest is Giles's discussion of how the Jackson State shootings have become buried in American institutional memory, allowing us to forget the forces that made them possible.

Valera T. Francis and Amy E. Wells, in their chapter "On Opposite Sides of the Track: New Orleans' Urban Universities in Black and White," provide a much-needed historical account of the evolution of New Orleans' two public urban universities, Southern University of New Orleans (SUNO) and the University of New Orleans (UNO). Given current events in New Orleans, theirs is a bittersweet study of the motives and forces behind the establishment of a segregated system of higher education in the South. These motives and forces have changed, of course, as a result of the civil rights gains, but they have not disappeared. What can only be called the "politics of Hurricane Katrina" illustrates that in New Orleans, and by extension many other urban areas in the South (and in the North), we are still very much on opposite side of the tracks.

In Chapter 9, Kristen Safier examines the impact of federal civil rights investigations over the past 25 years at Central State University in Ohio. She cogently describes the complicated relationships that public Black colleges have had with their state legislatures, describing a tight rope of sorts. Safier urges public Black college administration to establish strong relationships with federal agencies as they appear, according to her legal case study, to be more sympathetic to the needs and challenges of Black colleges than state governments.

Another public Black college that has had troubles in recent years is Texas Southern University. The institution has faced leadership challenges as well as a public critique pertaining to the quality of education the institution offers. However, very little is known, beyond anecdote, about student success at Texas Southern. "The Retention Planning Process at Texas Southern University: A Case Study," contributed by Jacqueline Fleming, Albert Tezeno, and Sylvia Zamora, aims to shed light on the retention of African American students at Texas Southern. Of note, each of these individuals is affiliated with the institution and as such, they offer a unique insider perspective. In their chapter, Fleming and colleagues discuss the persistent problem of low graduation rates within the Black college context. More importantly, they profile a successful retention effort at Texas Southern, giving hope to those Black colleges that are struggling to increase their graduation rates.

The third section of this edited volume focuses on subjects that are often considered *Taboo* within the larger Black community, and society at large. In Chapter 11, titled "Color and Class: The Promulgation of Elitist Attitudes at Black Colleges," Bianca Taylor explores one of the most taboo topics. She uses the Black college context as a lens through which to examine the manifestations of colorism and classism within the Black community. Drawing on the work of the sociologist E. Franklin Frazier, Taylor examines the ways in which education has been a main factor in promoting the Black elite, but more importantly, she connects this examination with color and class rifts, rarely discussed, but deeply embedded in African American culture.

"Not a Laughing Matter: The Portrayals of Black Colleges on Television," written by Adam Parrott-Sheffer, offers a fascinating glimpse into the depictions of Black colleges in television programming. Parrott-Sheffer explores positive, negative, and stereotypical (be they positive or negative) depictions of Black college students, faculty, and administrators, comparing representations on television with the scholarly literature on Black colleges. He asks us to consider what kind of impact negative images of Black colleges, such as those on *College Hill*, have on "real" Black colleges. Given the general public's lack of knowledge of the strengths of Black colleges, do these images have an even greater impact than those negative images of HWIs in the media?

The last chapter in this book, written by Nelwyn B. Moore, J. Kenneth Davidson Sr., and Robert Davis, is "Sexual Behavior Patterns and Sexual Risk-Taking among Women and Men at a Historically Black University." The authors explore the provocative topic of sexuality, examining the belief systems that students acquire at home and how these collide with new ideas and belief systems acquired in the college environment. Their findings tell us much about the role of religion in sexual decisions, students' sexual risk-taking behaviors, and the level of comfort that Black students have with the subject of sexuality. Perhaps the most important aspect of this work is the spotlight it puts on issues of sexuality within the Black college and Black community contexts—a spotlight that Moore and colleagues urge us to shine more brightly until we have a better sense of Black college student sexual practices.

Overall, we hope that the essays in this book push the reader to think beyond the status quo. We also hope that these chapters can be used as a jumping-off point for those interested in conducting research on Black colleges, especially more taboo subjects. Although, by and large, Black colleges face many of the same challenges as their historically White counterparts, because of racism (both historic and current forms), these institutions tend to garner greater criticism. Moreover, Black college failures are generalized to the whole lot of Black institutions, whereas their successes are rarely celebrated by the mainstream educational community, policymakers, and the media. It is vital that scholars with an ethic of care and fairness conduct research on Black colleges, highlighting their achievements and pointing out their challenges so that those with an interest and concern can work on addressing these challenges. It is from this perspective that we approached the writing of this book.

REFERENCES

Anderson, J. D. 1988. *The education of Blacks in the South*. Chapel Hill, NC: University of North Carolina Press.

American Association of University Professors. January–February, 1995. Historically Black colleges and universities: A future in the balance. *Academe* 81(1): 49–58.

Bennett, P., and Y. Xie. 2003. Revisiting racial differences in college attendance: The role of historically Black colleges and universities. *American Sociological Review* 68(4): 567–80.

Bridges, B., B. Cambridge, G. D. Kuh, and L. H. Leegwater. 2005. Student engagement at minority-serving institutions: Emerging lessons from the BEAMS project. *New Directions for Institutional Research* 125: 25–43.

Brown, M. C. 1999. *The quest to define collegiate desegregation: Black colleges, Title VI compliance, and post-Adams litigation*. Westport: CT: Greenwood Press.

———. 2001. Collegiate desegregation and the public Black college: A new policy mandate. *Journal of Higher Education* 72(1): 46–62.

Brown, M. C., and J. E. Davis. 2001. The historically Black college as social contract, social capital, and social equalizer. *Peabody Journal of Education* 76(1): 31–49.

Brown, M.C., and K. Freeman. 2004. *Black colleges: New perspectives on policy and practice*. Westport, CT: Praeger.

Brown, M. C., S. Donahoo, and R. Bertrand. 2001. The Black college and the quest for educational opportunity. *Urban Education* 36(5): 553–71.

Butchart, R. 1980. *Northern schools, Southern Blacks, and Reconstruction: Freedmen's education, 1862–1875.* Westport, CT: Greenwood.

Cohen, R. T. 2000. *The Black colleges of Atlanta.* Mount Pleasant, SC: Arcadia.

Fleming, J. 2001. The impact of a historically Black college on African American students: The case of LeMoyne-Owen College. *Urban Education* 36(5): 597–610.

Foster, L. 2001. The not-so-invisible professor: White faculty at the Black colleges. *Urban Education* 36(5): 611–29.

Freeman, K. 2002. Black colleges and college choice: Characteristics of students who chose HBCUs. *Review of Higher Education* 25(3): 349–58.

Freeman, K., and R. T. Cohen. 2001. Bridging the gap between economic development and cultural empowerment: HBCUs' challenges for the future. *Urban Education* 36(5): 585–96.

Gasman, M. 2007. "Swept Under the Rug?": A historiography of gender and Black colleges. *American Educational Research Journal* 44(4): 760–805.

———. 2007. *Envisioning Black colleges: A history of the United Negro College Fund.* Baltimore, MD: Johns Hopkins University Press.

Gasman, M. and S. Anderson-Thompkins. 2003. *Fund raising from Black college alumni: Successful strategies for supporting alma mater.* Washington, DC: Council for the Advancement and Support of Education.

Hall, B. 2005. When the majority is the minority: White graduate students' social adjustment at a historically Black university. *Journal of College Student Development* 46(1): 28–42.

Harper, S. 2004. Gender differences in student engagement among African American undergraduates at historically Black colleges and universities. *Journal of College Student Development* 45(3): 271–84.

Higher Education Act, 1965.

Jackson, C. 2003. *Historically Black colleges and universities: A reference handbook.* Santa Barbara, CA: ABC-CLIO.

Johnson, B. 2001. Faculty socialization: Lessons learned from urban Black colleges. *Urban Education* 36(5): 630–47.

Lewis, David L. 1994. *W. E. B. Du Bois. Biography of a Race, 1868–1919.* New York: Henry Holt.

Merisotis, J. P., and K. McCarthy. 2005. Retention and student success at minority-serving institutions. *New Directions for Institutional Research* 125: 45–58.

Minor, J. 2005. Faculty governance at historically Black colleges and universities. *Academe* 91(3): 34–7.

National Science Foundation. 2004. www.nsf.gov.

Perna, L. 2001. The contributions of historically Black colleges and universities to the preparation of African Americans for faculty careers. *Research in Higher Education* 42(3): 267–94.

Perna, L., V. Lundy Wagner, N. Drezner, M. Gasman, S. Yoon, E. Bose, and S. Gary. 2007. The contributions of historically Black colleges and universities to the preparation of African American women for STEM careers: A case study. Paper presented at the Association for the Study of Higher Education Annual Meeting.

Sissoko, M. 2005. Minority enrollment demand for higher education at historically Black colleges and universities from 1976 to 1998: An empirical analysis. *Journal of Higher Education* 76(2): 181–208.

Tabbye, C., A. Harris, D. Rivas, L. Helaire, and L. Green. 2004. Racial stereotypes and gender in context: African Americans at predominantly Black and predominantly White colleges. *A Journal of Research* 51(1/2): 1–16.

Watkins, W. 2001. *White architects of Black education: Ideology and power in America.* New York: Teachers College Press.

Wenglinsky, Harold. 1999. Historically Black Colleges and Universities: Their Aspirations and Accomplishments. Princeton, NJ: Educational Testing Service.

PART 1

TRIUMPHS

FOR *ALMA MATER* AND THE FUND: THE UNITED NEGRO COLLEGE FUND'S NATIONAL PRE-ALUMNI COUNCIL AND THE CREATION OF THE NEXT GENERATION OF DONORS

NOAH D. DREZNER

> I believe that these colleges [historically Black] form a legacy ever devoted to the high ideals for which they were grounded and the pre-alumni council should promulgate not only the tenets of higher education but should promote the highest in race relations as well.
>
> National Pre-Alumni Creed United Negro College Fund.

Member colleges of the United Negro College Fund (UNCF), like most other colleges and universities, require voluntary support in order to balance their budgets and achieve their institutional missions. While acknowledging the need for alumni support as a result of decreases in funding from federal and state governments, many private and public institutional advancement programs do not instill a culture of giving in their students, nor do they teach the importance of providing financial support to their *alma mater* after receiving their diploma (Dysart, 1989; Zusman, 1999). Nancy Dysart (1989) suggests that colleges may be forced to teach a culture of giving to their students in order to encourage future alumni participation in development campaigns. The UNCF, through its National Pre-Alumni Council (NPAC), aims to involve students in fund-raising in order to encourage their support after graduation. Alumni support accounts for the majority of voluntary giving to the general academy.

For instance, alumni provided 28% of the private donations to higher education in the fiscal year 2001 (American Association of Fundraising Counsel [AAFRC], 2003). However, only 12.2% of alumni solicited support to historically Black colleges and universities (HBCUs) in the same year (Voluntary Support of Education survey, 2005). Currently, fund-raising at colleges and universities is growing, resulting in increased competition. Voluntary support for higher education, from all sources, amounted to $24.2 billion in the fiscal year 2000, up from $7.56 billion (in constant 2000 dollars) in 1965 (AAFRC, 2005; Drezner, 2007). Fund-raising within higher education will only increase. As of March 2005 there were 49 campaigns, either announced or completed, with goals of raising over a billion dollars at institutions in the United States (Capital campaign status, 2005). John Lippincott, president of the Council for the Advancement and Support of Education (CASE), predicts that in the near future "one or more institutions will announce $5-billion campaigns . . . the next psychological threshold to cross" (Strout, 2005). Nearly 2,000 private institutions and more than 1,500 public two- and four-year colleges and universities compete with each other for the same donations. Duronion and Loession (1991, p. 1) described this trend as follows:

> The competition for private dollars, both within the field of higher education and throughout the entire nonprofit world, is more vigorous now than ever before. For some institutions, doing well in this competition is no less than a matter of survival. For all institutions, competing successfully for private support provides the money to ensure institutional growth and strength.

As a result of this competition for dollars, administrators are searching for all possible revenue sources to help meet budget requirements. Therefore, encouraging students to give and maintain their commitment and loyalty to their *alma mater* as alumni is very appealing to college administrators.

College and university advancement offices ponder over how to motivate donors and how to encourage alumni to make financial contributions to their *alma mater* on a regular basis. According to social psychologists it is possible to learn prosocial behavior, or voluntary actions that are carried out to benefit others (Eisenberg, 1982; Rushton, 1982; Schroeder et al., 1995). One example of prosocial behavior is alumni making voluntary donations to their *alma mater.*

Studies show that as a person ages, developmental and moral reasoning can evolve with regard to helping others (Cialdini and Kenrick, 1976; Bar-Tal, 1982; Eisenberg, 1982; Kohlberg, 1985; Schroeder et al., 1995). Young children offer help as a result of extrinsic motivation—being told to help, being punished, or being promised a gift or prize. Less tangible benefits, such as peer approval, are associated with adolescent motivation to help others. Adults reach a different stage, where intrinsic feelings motivate their prosocial behavior (Cialdini and Kenrick, 1976; Bar-Tal, 1982; Eisenberg, 1982; Kohlberg, 1985; Schroeder et al., 1995). Schroeder et al. (1995), drawing on previous research, suggest that prosocial behavior can be taught and learned. Numerous scholars believe that direct reinforcement as well as observation and discussion of altruism influence prosocial behavior (Ahammer and Murray, 1979; Grusec, 1982, 1991; Israel, 1978; Rushton, 1975, 1982; Moore and Eisenberg, 1994; Smith et al., 1979).

THE UNITED NEGRO COLLEGE FUND AND ITS NATIONAL PRE-ALUMNI COUNCIL

The UNCF, established in 1944, is a comprehensive fund-raising organization that solicits individuals, corporations, and foundations. The organization provides operating funds for its 39 private member HBCUs, scholarships and internships for students at hundreds of institutions, and faculty and administrative professional training. One extension of the UNCF is the NPAC, which is a student alumni association in which participants are engaged in fund-raising and alumni-relationship building for their institution, or in this case, their *alma mater* and the UNCF (Tipsord Todd, 1993). Recently, there has been research on the power of student alumni associations in engaging future alumni to help create a new generation of volunteers and donors at historically White institutions (HWIs) (Friedmann, 2003). However, this research neglects to look at Black colleges and universities and their unique position in having a community chest fund-raising arm—the UNCF. The UNCF's NPAC was established in 1958

> to stimulate interest and participation of students in the programs of the United Negro College Fund, to preserve and to further loyalty and fellowship between the member colleges and universities, assist in raising funds during the annual campaign and help them become better alumni while in school and upon graduation.
> (UNCF, n.d.)

More specifically, the NPAC identified six goals to further its purpose. Of them, three are directly related to connecting students to the fund-raising needs of the UNCF:

1. Encourage individual [student] contributions to the College Fund/UNCF.
2. Give maximum assistance in raising funds during the annual campaigns of the College Fund/UNCF and NPAC member institutions.
3. Stimulate awareness among college students of the need to become active members in their respective Pre-Alumni Councils (PACs).
   (Jefferson, 2004, p. 2)

Simply put, the NPAC's purpose is to instill a culture of giving in its members. The intention of this chapter is to explore the UNCF NPAC as a model of socialization and how it involves and teaches the next generation of alumni donors the importance of supporting their *alma mater* and the UNCF. To explore this socialization process, I conducted a case study of the UNCF NPAC (Creswell, 2003). I interviewed current NPAC student leaders and past and present professional advisors throughout the nation and reviewed organizational documents, including NPAC newsletters, presentations, reports, and plans, to inform my findings.

HISTORY AND CONTEXT

Then president of Tuskegee Institute, Fredrick D. Patterson founded the UNCF in 1944 (Gasman, 2007). The UNCF's mission was to provide assistance to Black college students through scholarships and to raise operating funds for its then 27 member colleges and universities. Over the past 62 years, individual, corporate, and foundation donors have contributed over $2.1 billion to the UNCF and helped more than 300,000 students earn undergraduate and graduate degrees (UNCF, 2005). The organization generated its funds from individuals through a number of avenues: alumni appeals, solicitation of celebrities, annual telethons, and student fund-raising through the NPAC system (Rucker, personal communication, November 8, 2005; Wanga, personal communication, November 10, 2005).

In 1946, two years after the founding of the UNCF, James E. Stamps, an alumnus of UNCF member college Fisk University (Nashville, TN), established the National Alumni Council (NAC). The NAC's purpose was to bring alumni together in partnership with the UNCF to serve the needs of private HBCUs. Twelve years later, in 1958, Walter Washington, a Tougaloo alumnus and then president of Utica Junior College (Utica, MS), created the NPAC under the NAC. Washington established the NPAC to foster loyalty between institutions, their alumni, and the UNCF (UNCF, n.d.). Over the last 47 years NPAC students have raised more than $3 million for the UNCF annual campaigns (Benedict College Division of Institutional Advancement, n.d.; Wanga, personal communication, November 10, 2005).

Student involvement in fund-raising activities creates a strong foundation for active alumni support after graduation. For instance, Friedmann (2003) found that involvement in a student alumni association has a significant, positive effect on alumni giving and on campaign participation, cumulative lifetime giving, and size of individual gifts.

In a 1981 interview about his involvement in the founding of the UNCF, James P. Brawley, then president of Clark College in Atlanta, Georgia (now Clark-Atlanta), understood the importance of instilling a culture of giving in undergraduates so that they would become donors as alumni. Brawley (1981, n.p.) believed,

> If you are going to develop responsive alumni you don't do it by talking to them when they are in their caps and gowns ready to go, and then expect them to respond by giving handsome gifts to the college . . . the need is to develop a systematic plan for the alumni to contribute and stimulate their interest through what is done while they are at the college for four years, and if you don't get a good response out of them during those four years, the chances are 99[%] that you won't get much of a response after they have gone.

Substantial literature exists that corroborates Brawley's notion and discusses the importance of engaging students in programming, in solicitations as student callers, and even as donors early in their careers at both two- and four-year institutions (Chewning, 1993; Kerns, 1986; Lynch, 1980; Nakada, 1993; Nayman et al., 1993; Purpura, 1980; Shanley, 1985; van Nostrand, 1999). Nayman et al. (1993, p. 90) suggest,

> Turning students into donors is a socialization process that involves orienting students to the notion of voluntary giving, actively engaging them in varied institutional advancement activities, and strategically timing program initiatives.

By socializing the students in this way, Nayman et al. find that students are more likely to participate in future fund-raising campaigns. It is this socialization process that frames this study.

BOND TO *ALMA MATER* AND THE UNCF

Successful fund-raising involves the donor feeling a personal connection to the organization to which he or she is giving (Mael and Ashforth, 1992). Student alumni associations, such as the UNCF's NPAC, are able to harness that identification and successfully solicit their members' support. Organizational identification, a part of social identity theory, is the act of an individual defining himself or herself by an organization. In the case of higher education saying "I am a student at . . . " or "I am an alumna of . . . " qualifies as organizational identification (Ashforth and Mael, 1989; Mael and Ashforth, 1992). Mael and Ashforth (1992, p. 104) suggest that college alumni conceptualize organizational identification because

> (1) college can be considered a "holographic organization" (Albert and Whetten, 1985), that is, one where members share common organization-wide identity and are less likely to experience competing demands from, say, department-level or occupational identities, and (2) since alumni constitute a particularly critical source of support for colleges, alumni identification is likely to strongly affect the welfare of their respective *alma maters*.

Mael and Ashforth (1992) also propose a relationship of organizational identification in which both aspects of the institution and individual feed into an alumnus' organizational identity, which then leads to an "organizational consequence" of supporting their *alma mater*. Using social identity theory as a basis, Mael and Ashforth predict that alumni identification with their *alma mater* corresponds to participation in gift campaigns and alumni relations events and encouraging others to attend the institution.

HBCUs perhaps have a better opportunity in creating the personal and organizational identity that Mael and Ashforth (1992) suggest. Gasman and Anderson-Thompkins (2003, pp. 37–8) find that

> for many Black-college alumni, the bond to alma mater is formed long before they arrive on campus—especially in the case of legacies . . . The college is "alma mater" in the truest sense because it nurtured them much like a mother and gave them skills that they might not get elsewhere in a White-dominated society. If nurtured and re-kindled regularly, the surrogate parent image can be beneficial to institutional fund raising; if neglected, it can be devastating to alumni giving.

It is this "bond to alma mater" and the UNCF that the NPAC wants to develop and enhance through stimulating the interest and participation of students enrolled at member institutions (Wanga, 2005).

Chase Gayden (2005, pp. 4–5), the southwestern regional director of NPAC and a junior at Tougaloo College, identifies the "bond to alma mater" and the UNCF as

> a lifetime of insurance for me. Owing to its support, I realize that education is the key to success, a bus to a brighter future for our people. Without education, there is little that a person can do-actually there is not a lot a person can do without an education. Through the UNCF I have been able to build my education, to strengthen it and to replenish it. We must continue to be knowledge seekers and we

must strive for a better life through education and support of the United Negro College Fund.

Another executive board member, Lavretta Moore, ties her bond to the UNCF and NPAC further to the African American experience:

> [The] Pre-Alumni [Council] is significant on so many levels because it plants the mental seed and serves as a gentle reminder that our academic achievements and educational advancements are possible only because of the blood, sweat, and tears shed by our ancestors. We, as African-Americans, as descendants of slaves and share-croppers, as a people that once were not allowed to attend "other" institutions of higher learning, owe it to ourselves and one another to maintain and sustain our own; at one point, it was all we had!
>
> (Moore, 2005, p. 5)

She continues, noting that her involvement in the NPAC is as a scholarship recipient, which is "a financial blessing," and sees her participation as "one of the only ways I am able to reinvest a mere portion of what the organization has vested in me" (Moore, 2005, p. 5).

INVOLVEMENT LEADING TO EDUCATION

Students involved in the NPAC fund-raising efforts are likely to give and participate at a greater rate than their peers in postgraduation campaigns by the UNCF and their individual institutions (Nayman et al., 1993). Friedmann (2003, p. 80) notes that student alumni association advisors believed that "students involved in their organizations develop greater prosocial behavior and increased intrinsic motivation toward their institutions as alumni compared to members of the general student population." By participating in fund-raising activities while in school, students "have a greater awareness and understanding of the significance of private giving to the institutions" than their peers (Friedmann, 2003, p. 80).

Friedman's findings were confirmed by the current president of the NPAC's executive board. She believes that "involvement in NPAC opens pre-alumni's eyes to the work of the UNCF and how their education and their institution have benefited from the UNCF" (Wanga, personal communication, November 10, 2005). This idea is corroborated by other NPAC executive board members. Gayden (2005, p. 4) believes that his involvement in NPAC as the executive board's southwestern regional director and member of the Tougaloo College Pre-Alumni Council not only taught him about the need for fund-raising for the UNCF, but it prepares

> students to become effective alumni under the guidance of dedicated alumni. It is important that we learn as pre-alumni about our responsibility to give back to our communities and schools, in hopes of allowing other young people to have the same opportunities which we have been privileged enough to receive.

The notion of opening students' eyes to the needs of the UNCF and their *alma maters* is important. This education process and demystification of Black college needs, done prior to graduation, can help increase future giving. There is a common misconception on the part of many HBCU alumni that the institution does not need support (Gasman and Anderson-Thompkins, 2003). Many alumni simply believe that their *alma mater* is strongly funded by the federal government and supplemented by corporations and foundations and therefore not in need of alumni support (Gasman and Anderson-Thompkins, 2003). By engaging students in organizations, such as NPAC, which encourage philanthropy and educate them about the UNCF's needs, the participants are more likely to take part in annual campaigns (Friedmann, 2003).

THE USE OF BENEFITS TO ENCOURAGE PARTICIPATION

Colleges and universities regularly solicit their alumni for donations, using intrinsic and extrinsic benefits as motivation (Worth, 2002). Extrinsic motivations may include small gifts, invitations to campus activities, listing of names in annual reports that are widely read by their peers, membership in giving societies, or the belief that alumni participation and dollars increase their *alma mater*'s reputation and therefore the value of their own degrees. In contrast, intrinsic motivations for giving include alumni giving to a scholarship fund that helps others attend college (Harbaugh, 1998).

The NPAC uses both extrinsic and intrinsic motivations to raise funds from its members. The most extrinsic motivation that the NPAC uses is open only to women—the Miss UNCF competition. Rather than a beauty contest, the Miss UNCF title recognizes a young woman's fund-raising ability on behalf of the UNCF. Each UNCF member college can send a Miss UNCF candidate to the annual NPAC conference for the title. Local Miss UNCFs are determined on the basis of the amount of money that the individual pre-alumna raised in the school's UNCF Drive, typically held in the fall semester. The woman who raises the largest amount (alternating yearly between aggregate amount raised and amount raised per capita student on her campus) across all 39 member institutions captures the Miss National UNCF title at the annual joint national alumni/pre-alumni conference. The "coronation" of Miss UNCF is the culmination of the conference. The ceremony is choreographed with the pomp and circumstance of a Miss America pageant. With the title comes a UNCF scholarship, a position on the NPAC executive board, and a seat on the NAC board of directors, in which Miss UNCF acts as the liaison and spokeswoman of her fellow undergraduates to the UNCF and its sponsoring organizations (Wanga, personal communication, November 10, 2005).

The Miss UNCF program has been very effective. The 2006 winner, Wiley College student Natasha Jenkins, raised $50,000 for the UNCF (UNCF, 2006). However, this extrinsic motivation is geared only to female students. Men, not being allowed to compete for an equivalent title such as Mr. UNCF, are likely not as motivated as women to raise funds for the UNCF without the prestige, influence, and scholarship incentives that their female counterparts are given. By expanding such a successful program to men, perhaps the UNCF and the students' *alma mater* will benefit in raising more money not only in the short term, but in the long term as well, by creating another population of engaged young alumni with a stronger commitment to and relationship with the UNCF and their individual institution.[1]

On an intrinsic level, many pre-alumni council members solicit others for the annual fund-raising drives and give to their *alma mater* and the UNCF in order to make sure that other Black students receive scholarships and have the opportunity to attend HBCUs. The vice president of the NPAC, Wilberforce University junior Tatum Rucker, is not on a UNCF scholarship. However, she volunteers, fundraises, and gives simply because

> for many years the United Negro College Fund has provided opportunities for African Americans to receive a higher education. As a college student and being knowledgeable of the hardships of financing an education, I wanted to be involved in providing scholarship opportunities for my peers. Although I am not on a UNCF Scholarship, I still recognize the significance of the UNCF/College Fund. Being involved in the NPAC has provided me with that opportunity.
>
> (Rucker, personal communication, November 8, 2005)

On one level, this sense of "giving back" is consistent with other student alumni associations in which students report a "feeling of reciprocity, giving back to the [institution] that helped them" (Friedmann, 2003, p. 109). However, this rationale for involvement is even more interesting when looking at giving as a statement of racial uplift. Gasman and Anderson-Thompkins (2003) bring forward historical evidence and current research that support the notion that Black donors are motivated to give or give back in order to promulgate racial uplift. Further, these authors find that Black college alumni often give "to continue the legacy of . . . Black college[s]" (2003, p. 38). Robiaun Charles, a former Pre-Alumni Council advisor and advancement officer at Johnson C. Smith University (Charlotte, NC), believes that it is the students who perceive the UNCF as something "racially uplifting" and use those messages in their activities rather than finding messages of racial uplift within official UNCF materials (Charles, personal communication, March 13, 2006).

When asked "Why is the UNCF important today?" Wanga (UNCF, 2006, p. 2) reflects this idea of being involved and giving back as a mechanism for future racial uplift:

> UNCF is important today because it gives hope to the hopeless, life to the lifeless, healing to the broken-hearted and opportunity to the disadvantaged. Society has labeled these students as uneducable and lazy, however UNCF has illuminated its torch of opportunity and recognized these same students as young bright minds.

In this statement, Wanga's emotion goes beyond the typical feeling and donor motivation—showing how she believes that the UNCF, and therefore Black colleges, are instrumental in uplifting the race.

ENCOURAGING OTHER FORMS OF PHILANTHROPY

Kang (2005) suggests that participating in other forms of philanthropy, such as devoting time to community service while a student, also has a positive effect on future giving. Likewise, Rushton (1982) believes students involved in a group that encourages various forms of philanthropy (monetary giving, service, etc.) have an increased socialization to the organization and stronger identity, which increases their prosocial behaviors. Building off of this theory, the NPAC requires its member councils at each institution to engage in a service project within their campuses' home community—often supporting Black organizations (Rucker, personal communication, November 8, 2005). The NPAC suggests that local PACs engage in service to "spread the mission of [sic] UNCF and promote unity between your PAC, campus, and community" (Rucker, 2005, p. 1).

The NPAC executive board promotes community service programs such as (1) adopt-a-class, where participants work with a primary or secondary class and implement activities that have an educational and historical component; (2) health disparities forums, where students choose a health issue that affects their community and invite experts to share information on the topic with fellow students and the local community; and (3) high school visitation days, in which local secondary students are invited to the campus for a college tour and information about the UNCF and other member institutions (Rucker, 2005). The high school visitation days project grows out of the goal to "assist [the National Alumni Council] in its aim to encourage young people to attend College Fund/UNCF member institutions" (Jefferson, 2004, p. 2). In addition to these projects, the National Miss UNCF designs and coordinates the implementation of a national service project that culminates in the annual NAC/NPAC conference. By engaging in large community service projects such as these, not only does the PAC promote itself and its mission, but gives the participants another avenue to teach the members about the importance of helping others, and more specifically, the Black community.

IMPLICATIONS ON FUTURE ALUMNI GIVING

Building upon Atchley's (1989) continuity theory, Lindahl and Winship (1992) and Okunade and Justice (1991) find that past giving behaviors are correlated with current and future giving practices. These findings suggest that by enhancing students' prosocial behavior through instilling a culture of philanthropy, colleges can establish relationships that will continue long after graduation. Research shows that young alumni giving, even in small amounts, has the potential of having great effects on lifetime giving (Lindahl and Winship, 1992; Monks, 2003; Nayman et al., 1993; Okunade and Justice, 1991).

Texas College senior and the current president of the NPAC, Carolina Wanga, believes that alumni who are involved in their PAC are more active in the UNCF and their institution after graduation. She has seen the most involved students become active young alumni in the UNCF NAC and their institution's regional alumni councils or interalumni councils. Wanga asserts that one of the reasons for this transformation is the students' observation of dynamic alumni involved in activities such as the annual NAC/NPAC conference (personal communication, November 10, 2005).

Understanding that the NPAC encourages students to participate in the monetary support of the UNCF and their *alma mater* and in the service to their community, we can expect, when applying the continuity theory, to observe that after graduation those involved in the NPAC will be continually engaged in the support of the UNCF and their institutions.

NOTE

1. A few UNCF schools have local "Mr. UNCF" contests but these are not recognized at the national level.

REFERENCES

Ahammer, I. M., and J. P. Murray. 1979. Kindness in the kindergarten: The relative influence of role playing and prosocial television in facilitating altruism. *International Journal of Behavioral Development* 2: 133–57.

American Association of Fundraising Counsel. 2003. *Giving USA 2003: The annual report on philanthropy for the year 2002.* New York: Author.

American Association of Fundraising Counsel. 2005. *Giving USA 2005: The annual report on philanthropy for the year 2004.* New York: Author.

Ashforth, B. E., and F. Mael. 1989. Social identity theory and the organization. *Academy of Management Review* 14: 20–39.

Atchley, R. C. 1989. A continuity theory of normal aging. *Gerontologist* 29(2): 183–90.

Bar-Tal, D. 1982. Sequential development of helping behavior: A cognitive-learning approach. *Developmental Review* 2: 101–24.

Benedict College Division of Institutional Advancement. n.d. The National Pre-Alumni Council Organization. Retrieved September 21, 2005, from http://www.benedict.edu/divisions/insadv/alumni_affairs/alumni/bc_pre_alumni_council.html.

Brawley, J. P. 1981. Oral history interview by Marcia Goodson, Columbia University Oral History Collection, New York.

Capital campaign status. 2005. *The Chronicle of Higher Education.* Retrieved April 5, 2005, from http://www.chronicle.com/prm/weekly/campaigns/.

Chewning, P. B. 1993. The ultimate goal: Installing the volunteer and philanthropic ethic. In *Student advancement programs: Shaping tomorrow's leaders today,* ed. Barbara Tipsord Todd. Washington, DC: Council for the Advancement and Support of Education.

Cialdini, R. B., and D. T. Kenrick. 1976. Altruism as hedonism: A social development perspective on the relationship of negative mood state helping. *Journal of Personality and Social Psychology* 34: 907–14.

Creswell, J. W. 2003. *Research design: Qualitative, quantitative, and mixed methods approaches.* Thousand Oaks: Sage.

Drezner, N. D. 2005. Advancing Gallaudet: Alumni support for the nation's university for the deaf and hard-of-hearing and its similarities to Black colleges and universities. *International Journal of Educational Advancement* 5(4): 301–15.

Drezner, N. D. 2007. Recessions & tax-cuts: The impact of economic cycles on individual giving, philanthropy, and higher education. *International Journal of Educational Advancement* 6(4): 289–305.

Duronion, M. A., and B. A. Loession. 1991. *Effective fund raising in higher education.* San Francisco, CA: Jossey-Bass.

Dysart, N. M. 1989. Alumni-in-resident: Programs for students. In *Handbook for alumni administration,* ed. C. H. Webb. New York: Macmillan.

Eisenberg, N., ed. 1982. *The development of prosocial behavior.* New York: Academic Press.

Friedmann, A. S. 2003. Building communities of participation through student advancement programs: A first step towards relationship fund raising. PhD dissertation, The College of William and Mary in Virginia, 2003).

Gasman, M., and S. Anderson-Thompkins. 2003. *Fund raising from Black college alumni: Successful strategies for supporting alma mater.* Washington, DC: Council for the Advancement and Support of Education.

Gasman, M. 2007. *Envisioning Black colleges: A history of the United Negro College Fund.* Baltimore, MD: Johns Hopkins University Press.

Gayden, C. 2005, June 1. Introduction of the 2005–2006 NPAC Board. *The Torch.* Retrieved on September 30, 2005, from http://www.texascollege.edu/pdf/Torch_Newsletter_Summer.pdf.

Grusec, J. E. 1982. The Socialization of altruism. In *The development of prosocial behavior,* ed. N. Eisenberg, 139–66. New York: Academic Press.

———. 1991. The socialization of empathy. In *Review of personality and social psychology: vol. 12. Prosocial behavior,* ed. M. S. Clark, 9–33. Newbury Park, CA: Sage.

Harbaugh, W. 1998. The prestige motive for making charitable transfers. *American Economics Review Papers and Proceedings* 88(2): 277–82.

Israel, A. C. 1978. Some thoughts on the correspondence between saying and doing. *Journal of Applied Behavior Analysis* 11: 271–76.

Kang, Chul-Hee. 2005, February. *An exploration on individual giving and volunteering in Korea.* Paper presented at colloquium of the University of Pennsylvania School of Social Work, Philadelphia.

Kerns, J. R. 1986, June. *Two-year college alumni programs into the 1990s.* Paper presented at the National workshop on two-year college programs, Junior and community college institute, Washington, DC.

Kohlberg, L. 1985. *The psychology of moral development.* San Francisco, CA: Harper & Row.

Jefferson, Z. 2004. The constitution and by-laws of the National Pre-Alumni Council of the UNCF, Inc. United Negro College Fund Papers, Library of Congress, Washington, DC.

Lindahl, W., and C. Winship. 1992. Predictive models for annual fundraising and major gift fundraising. *Nonprofit Management and Leadership* 3: 43–64.

Lynch, H. G. 1980. The young alumnus: An enduring strength. In *Building your alumni programs: The best of CASE Currents,* ed. V. Carter and P. A. Alberger. Washington, DC: Council for the Advancement and Support of Education.

Mael, F., and B. E. Ashforth. 1992. Alumni and their alma mater: A partial test of the reformulated model of organizational identification. *Journal of Organizational Behavior* 13(2): 103–23.

Monks, J. 2003. Patterns of giving to one's alma mater among young graduates from selective institutions. *Economics of Education Review* 22: 121–30.

Moore, B. S., and N. Eisenberg. 1994. The development of altruism. *Annals of Child Development* 1: 107–74.

Moore, L. 2005, June 1. Introduction of the 2005–2006 NPAC Board. *The Torch.* Retrieved on September 30, 2005, from http://www.texascollege.edu/pdf/Torch_Newsletter_Summer.pdf.

Nakada, L. H. 1993. Student interns: Cultivating the next generation of advancement professionals. In *Student advancement programs: Shaping tomorrow's leaders today,* ed. Barbara Tipsord Todd. Washington, DC: Council for the Advancement and Support of Education.

Nayman, R. L., H. R. Gianneschi, and J. M. Mandel. 1993, Fall. Turning students into alumni donors. In *New roles in educational fundraising and institutional advancement,* ed. M. C. Terrell and J. A. Gold, 63; 85–94. San Francisco, CA: Jossey-Bass.

Okunade, A., and S. Justice. 1991. Micropanel estimates of the life-cycle hypothesis with respect to alumni donations. In *Proceedings of the Business and Economics Statistical Section of the American Statistical Association,* 298–305. USA: American Statistical Association.

Purpura, M. 1980. Building the alumni habit. In *Building your alumni programs: The best of CASE Currents,* ed. V. Carter and P. A. Alberger. Washington, DC: Council for the Advancement and Support of Education.

Rucker, T. 2005. United Negro College Fund Pre-Alumni Council programs handbook. Unpublished document.

Rushton, J .P. 1975. Generosity in children: Immediate and long term effects of modeling, preaching, and moral judgment. *Journal of Personality and Social Psychology* 31: 459–66.

———. 1982. Social learning theory and the development of prosocial behavior. In *The development of prosocial behavior,* ed. N. Eisenberg. New York: Academic Press.

Schroeder, D. A., L. A. Penner, J. F. Dovidio, and J. A. Piliavin. 1995. *The psychology of helping and altruism: Problems and puzzles.* New York: McGraw-Hill.

Shanley, M. G. 1985. Student, faculty and staff involvement in institutional advancement: University of South Carolina. *Carolina View* 1: 40–3.

Smith, C. L., D. M. Gelfand, D. P. Hartmann, and M. E. P. Partlow. 1979. Children's casual attributions regarding help giving. *Child Development* 50: 203–10.

Strout, E. 2005, January 7. Fund raising: The big gifts had better get bigger. *Chronicle of Higher Education* 51(18): 7.

Tipsord Todd, B., ed. 1993. *Student advancement programs: Shaping tomorrow's alumni leaders today.* Washington, DC: CASE Books.

United Negro College Fund. n.d. Facts about the National Pre-Alumni Council (NPAC) and Pre-Alumni Councils (PACs). Retrieved September 12, 2005, from http://www.uncf.org/alumni/npacfacts.asp.

United Negro College Fund. 2005. About us. Retrieved September 12, 2005, from http://uncf.org/aboutus/index.asp.

United Negro College Fund. 2006. Taking it personally: UNCF students raise funds to support their collective education. Retrieved May 1, 2006, from http://www.uncf.org/webfeature/archives/webfeature_03142005_MissUNCF.asp.

van Nostrand, I. 1999. Young alumni programming. In *Alumni relations: A newcomer's guide to success,* ed. John A. Feudo. Washington, DC: Council for the Advancement and Support of Education.

Wanga, C. 2005. UNCF/NPAC 2005 initiatives: Strengthening the fabric of the NPAC, one fabric at a time. Presentation to executive board meeting.

Worth, M. J. 2002. *New strategies for educational fund raising.* Westport, CT: Oryx.

Zusman, A. 1999. Issues facing higher education in the twenty-first century. In *American higher education in the twenty-first century: Social, political, and economic challenges,* ed. P. Altbach, Robert O. Berdall, and Patricia J. Gumport. Baltimore, MD: Johns Hopkins University Press.

CHAPTER 2

ON FIRM FOUNDATIONS: AFRICAN AMERICAN BLACK COLLEGE GRADUATES AND THEIR DOCTORAL STUDENT DEVELOPMENT IN THE IVY LEAGUE

PAMELA FELDER THOMPSON

> Hence, it is the HBCU, with 20% total national enrollment of African American students but with a one-third national graduation rate for this same population, that will ensure the continued development of future academicians and leaders.
> Bonner and Evans, 2004, p. 5.

As a researcher in the area of graduate student development, I am often intrigued by the experiences of students who travail the doctoral process successfully, especially those who have received degrees from historically Black colleges and universities (HBCUs). As both an African American doctoral degree holder and a baccalaureate degree recipient from a historically Black university, I am familiar with both experiences. I have also listened to the experiences of African American mentors and colleagues who have endured similar experiences and shared with me their stories of success. I find these stories to be remarkably informative. These experiences not only exemplify the will and determination needed to achieve academic success, but they also provide lessons of endurance associated with academic persistence, and can inform the higher education community about success at the doctoral level.

The doctoral process, characterized as a novice graduate student transforming into a scholarly expert in research and teaching in a particular field of study, is an experience wrought with challenges. In fact, researchers have suggested that many students make this transition with little formal guidance (Boote and Beile, 2005; Margolis and Romero, 1998). Based on this premise, the attainment of a doctoral degree is a feat worthy of further examination and a student's undergraduate experience can inform researchers about doctoral program success.

According to previous research, a significant factor in examining retention at the undergraduate level is the understanding of the student experience from a cultural perspective (Tinto, 1993; Kerlin, 1995). Similarly, this holds true for the doctoral process as well. In fact, several studies indicate the need to understand doctoral student development especially as it relates to the experiences of specific cultural groups and institutions (Tinto, 1993; Davidson and Foster-Johnson, 2001; Kerlin, 1995; Thompson, 2005). While the literature on this topic is scant in general, there has been a growing number of researchers examining the importance of historically Black colleges and their contribution to the development of African American doctoral students (Willie, Grady, and Hope, 1991; Blackwell, 1987; Allen, 1992).

Largely, this scholarship related to African American doctoral students documents the critical importance of HBCUs and serves to refute previous claims that Black colleges should be eliminated because of their alleged role in facilitating segregation within higher education (Perna, 2001). Historically, these institutions have made great strides both in serving and educating African Americans. Gasman and Anderson-Thompkins (2003) support this notion in their assessment of Black college alumni by describing not only the essential role of these institutions in educating the African American population, but their vital contribution to American society in general. Moreover, Patton and Bonner (2001) assert that HBCUs have "provided immeasurable benefits by way of student leadership and social development" (p. 18). One wonders where the African American doctorate would be without the existence of HBCUs.

A study by Garibaldi (1997) highlights the contribution of HBCUs to the African American doctoral degree production in the United States during the 1990s, reporting that doctoral recipients between 1991 and 1995 indicated that 15 of the top 20 institutions that awarded these graduates their bachelor's degrees were HBCUs. According to his research, the top three producers of these graduates were Howard University (136), Spelman College (78), and Hampton University (69) (Garibaldi, 1997). Moreover, he found that out of the 5,284 African Americans who received their doctoral degrees between 1991 and 1995, 1,122 received their bachelor's degrees from HBCUs (1997). More recently, the Survey of Earned Doctorates (2003) captured statistics of the baccalaureate-granting institutions of racial/minority doctorate recipients in 2003. It highlighted that out of the top 20 institutions that produced 1,579 Black doctoral recipients, nine of the top ten were HBCUs.

Consequently, as the attainment of the doctoral degree is a means to increasing the number of minority faculty, the current status of minority faculty and the disproportionately low numbers of minority doctoral degree completers have prompted some retention and attrition experts to closely examine the experiences of students who do manage to successfully complete the doctoral process (Thompson, 2005; Taylor and Antony, 2000).

In addition, within the last few decades there has been a growing trend of researchers interested in the production of minority doctoral recipients (Taylor and Antony, 2000).

For example, in her work, "Black Women in Academe," Moses (1989) explores the subtle and constant racial and gender stereotypes that Black female faculty members endure as they struggle to incorporate their professional identities into the mission and goals of many historically White institutions (HWIs). As HBCUs continue to make important strides in developing African American leadership, their significant involvement in producing future doctoral degree recipients can serve as a model for student success. This chapter examines that issue more closely by conceptualizing the factors critical to the success of African American doctoral students and the contribution of HBCUs to this success. In particular, the chapter explores the experiences, belief systems, and behaviors of successful doctoral graduates from an Ivy League university who have attended HBCUs at the undergraduate level. The major questions that guide this study are as follows:

1. How have HBCUs influenced African American doctoral persistence within the Ivy League?
2. How have HBCUs facilitated the socialization of African American doctoral students within the Ivy League?
3. What are the belief systems and behaviors of successful African American doctoral students within the Ivy League who have attended HBCUs?

REVIEW OF LITERATURE

In his work, "Baccalaureate College of Origin of Black Doctorate Recipients," Brazziel (1983) found that HBCUs have had notable success in preparing their graduates for doctoral study. He asserts that the experiential characteristics of individuals who attend these institutions are a testament to the history and tradition of the graduate preparation legacy within these institutions. As with many Ivy League colleges and universities, Brazziel suggests that familial legacy seems to be a tie that binds success within the culture of many HBCUs. Likewise, in her analysis of the baccalaureate origins of successful African American women, Wolf-Wendel (1998) emphasized that historically Black women's colleges produced the largest proportion of successful African American women. She measures success by examining the baccalaureate origins of women who received their bachelor's degrees after 1965 from a U.S. college or university and who subsequently attained doctoral degrees between 1975 and 1991. This study continues to affirm the existence of HBCUs as the productivity ratios for these institutions showed that graduates were 47 times more likely to receive their undergraduate degrees from these institutions than HWIs. Comparatively, the ratios were six times greater for Black women's colleges than for historically Black coeducational institutions. Wolf-Wendel (1998, p. 143) further explained that the baccalaureate origin studies are especially "powerful" when considering ethnicity. Her work affirms Astin's (1962) belief that controlling for incoming student characteristics is significant to the institutional output disparities of PhDs and that self-selection evades the parameters of baccalaureate origin studies.

The HBCU experience for many students is often an odyssey that is life altering (Allen, 1992; Bonner and Evans, 2004; Thompson, 2005). As a graduate of a historically Black state institution, I can relate to students who have received their baccalaureate degrees from these colleges. I often describe them as being academically rigorous and socially conscious environments that challenge and shape students' intellect as well as their social, political, and spiritual lives. Many of my friends and colleagues who attended HBCUs are often still well connected with faculty, administrators, and classmates long after they graduate. Of course, this type of camaraderie and commitment is often characteristic of the relationships most people have with their undergraduate institutions. However, what makes the HBCU experience unique is the historical purpose and mission that guides the practices and policy making decisions that shape the preparatory development of its students.

The educational experience at an HBCU is often holistic, with the focus of educating the student both inside and outside of the classroom, a process that fosters cognitive growth, assertiveness, academic success, satisfaction, and general social-psychological adjustment (Allen, 1992). Students are academically educated with a formal curriculum. However, there is also an informal curriculum that students encounter in the HBCU environment that is critically important to their educational experience. Leading the informal and formal curriculums at these institutions are the HBCU faculty and administrators who are committed to serving underrepresented populations in higher education. Their commitment serves to promote the development of strong student-institutional relationships that lead to student retention (Allen, 1992; Palmer and Gasman, 2008).

In studying Black students' participation at predominantly Black and White institutions, Hughes (1987) suggests the importance of HBCUs and their role in supporting the adjustment and socialization of their students. She identifies five areas of development critical to Black undergraduate student success. They include (1) gender differences in Black students' college adjustment, (2) Black mental health on campus, (3) individuation and independence, (4) spiritual success, and (5) affiliation needs.

With regard to the gender differences, Hughes (1987) found that Black men and Black women were socially and culturally happiest at historically Black institutions. She also reports that Black students perceived the quality of their intellectual development to be comparable to that received at an HWI. Similar results were noted in examining the mental health of the study participants. Black students reported less stress and alienation at predominantly Black campuses. These factors can be critical to the development of a student's identity. This is also evident when Hughes discusses the evolution of a Black student's individuality and independence at a Black institution. Inherent to this development is support and encouragement from the "family, the extended family, friends and the home community" (Hughes, 1987, p. 540). Hughes further explained that this contact signified a respect for parental authority and upheld cultural meaning that promotes familial unity and strength. In addition to maintaining familial connections, Hughes (1987) sought to understand how spirituality contributed to student adjustment among her study participants. She found that reliance on spirituality was equally important to Black students at historically Black colleges and HWIs. And lastly, her study indicates that affiliation with the institution can contribute greatly to student adjustment. For example, Hughes (1987, p. 541) states, "The pattern of affiliation for Black students on Black campuses was more service oriented, whereas for Blacks on predominantly White campuses such affiliations resulted from social isolation experienced."

Hughes's (1987) study did not specifically address how HBCUs prepare African Americans for the doctoral process within the Ivy League. However, the study does support Allen's (1992) perspective of African Americans at the undergraduate level and the facilitation of their educational aspirations. Allen (1992, p. 40) found that "the educational goals of Black students are acted out in specific social environments that influence not only their ambitions, but also the possibility that they will realize their goals." His study draws a connection to the work of Tinto (1993), who asserts the importance of environment in social integration as a condition of student success. However, Allen notes that Black students are often able to succeed even when there are low levels of social integration that result in alienation and isolation particularly on predominately White campuses. Historical legacies of racial and cultural exclusion anchor barriers to academic achievement that are prevalent within these institutions (Wilson, 1988). While his study does not directly address issues of student development at the doctoral level, it does highlight factors significant to undergraduate African American achievement.

Tinto's (1993) Model of Doctoral Student Persistence outlines three general stages of the doctoral process. They include (1) transition to membership in the graduate community in the first year, (2) attaining candidacy through development of competence, and (3) active research. Traditionally, HBCUs have done a tremendous job in creating environments that foster many of the factors set forth in the doctoral student developmental models. For example, HBCUs have prepared students to overcome many of the barriers associated with the first-year doctoral experience (Tinto, 1993; Thompson, 2005; www.phdcompletion.org; www.preparing-faculty.org). For the most part, this is accomplished by exposing undergraduate and graduate students to activities that are specific to the doctoral process. These activities include introducing students to the concept of conducting high-quality research that merit publication, presenting this research within the larger educational community, and developing those research ideas to garner grant support. In addition, students are introduced to effective teaching and evaluation methods. And, students are introduced to faculty members (potential colleagues) in their field of expertise to promote collaboration and further exposure to the academic community. Moreover, the socialization experienced by these students in the HBCU environment prior to entering doctoral study serves as a firm foundation to managing the complex dynamics of an Ivy League environment.

RESEARCH METHOD

To further explore the impact of the HBCU undergraduate experience on the African American doctoral experience in the Ivy League, I conducted a study that assessed the African American doctoral experience, interviewing 11 doctoral degree completers who received their degrees from a School of Education at an Ivy League institution. Out of the 11 participants, 4 attended an HBCU as undergraduates. With little previous research guiding this study, qualitative methods were used to better understand the socialization experiences of these scholars who attended Ivy League institutions in an effort to contribute this knowledge to the existing body of literature. By way of personal alumni accounts, this study identified specific doctoral program preparation characteristics of the HBCU undergraduate experience.

A semi-structured interview protocol was used to gain an understanding of the participants' experiences. Some significant findings that evolved in the study were responses about the academic experiences of these students and the role of social networks at HBCUs. The participants shared reflections about both their academic and social preparatory experiences at these institutions.

FINDINGS

To illustrate how Black colleges are an integral factor in the experiences of African American doctoral students within the Ivy League several responses from doctoral degree completers who have received their doctoral degrees from an Ivy League institution but have received their baccalaureate degrees from three HBCUs (Howard University, Morehouse College, and Hampton University) are offered in the following pages. These quotes have been selected as they offer personal accounts of students' doctoral experiences and how their HBCU undergraduate education prepared them for degree completion. The names of the interviewees have been changed in the interest of confidentiality.

JAYLA (HOWARD UNIVERSITY)

Jayla, a 30-year-old graduate of Howard University, is a scholar who specializes in the area of reading, writing, and literacy. Her work involves using the media as an intervention to foster literacy interest among inner city disengaged teenage readers. Jayla believes that her experience in her PhD program was part of a higher calling and part of her life's purpose to continue her goals of improving literacy among young people. In fact, when I asked her why she chose to pursue her doctoral degree at an Ivy League institution, she replied, "I prayed on coming here . . . I never applied to any other doctoral program . . . I felt called to be here." When she mentioned this to me I embraced it as a familiar notion that many African Americans uphold the idea that one's education is more than self-improvement; it is a way to contribute to the improvement of community. This is an ideal prevalent in the organizational culture at most HBCUs; honored deeply by faculty, administration, and students. For example, in her interview Jayla discussed how this ideal was promoted on Howard's campus when she was a student there. She stated,

> Howard University is very steeped in a spiritual tradition that goes back into the antebellum South where prayer and camaraderie is valued on campus and I learned to pray . . . and get in line with other students who were about business and were about enlightenment and change. And, be serious about that and not flippant.
> (J. Capleton, personal communication, August 15, 2005)

Furthermore, Jayla explained how her prayer life and Howard's spiritual tradition was significant in preparing her for challenges beyond the undergraduate experience. She noted,

> I incorporated that kind of tenacity and hopefulness as a way of life and not just sort of the undergraduate experience to recollect and reminisce on but we were cultivating ourselves as adults and we were being cultivated as adults.
> (J. Capleton, personal communication, August 15, 2005)

While Howard's spiritual tradition was an important persistence factor in Jayla's undergraduate socialization, the issue of personal agency was an emerging theme in her interview responses as well. As noted earlier in the literature section, most doctoral students need to make the leap from student to self-directed researcher and scholar, and do so often with very little direction. According to Jayla, her experience at Howard prepared her for this challenge. She mentioned,

> Howard gave me a wealth of experience in speaking for myself; in practicing agency and autonomy even though it was a very strong community and familial space, the expectation was that as a person, as one individual contributing to the whole that you would be your best advocate. And, I had to learn to knock on doors, ask for what I want, how to stand my ground, argue my case, and seek opportunities instead of just standing and waiting for them to come to me.
>
> (J. Capleton, personal communication, August 15, 2005)

Jayla also described how Howard's rigorous academic environment prepared her for the academic experience in her Ivy League doctoral program. As a Research I institution, Howard has a reputation for staying on the cutting edge with its approach to scholarly training and a no-nonsense philosophy in integrating research with the political and social actions on campus in the community (www.phdcompletion.org). Jayla shared her perspective about her academic experience at Howard: "Howard was academically rigorous. There is always a lot to do and expectations were very high because the legacy at Howard is very rich" (J. Capleton, personal communication, August 15, 2005). Similarly, her learning experiences outside of the classroom supported her rich academic experience. In particular, she notes how being exposed to people from different backgrounds helped to broaden her horizons:

> I had a wealth of experience with different African people within the Diaspora, not only from America but from the Caribbean, Brazil, continent of Africa, and a wealth of heritages from different socio-economic backgrounds and class.
>
> (J. Capleton, personal communication, August 15, 2005)

Jayla's comments demonstrate evidence that her commitment to attaining a doctoral degree was grounded in her personal dedication to connecting her academic achievements with a spiritual purpose for her life. At Howard, she was able to develop this purpose within an environment that not only supported her vision of commitment, but provided a social network where that purpose was affirmed and groomed in ways that it could manifest in her professional career objectives. Jayla was able to align herself with individuals who held similar personal aspirations and used spirituality as a means of grounding their development.

As a member of the Howard community, Jayla was committed to being personally responsible to uphold her notions about social change. Howard's environment supported students engaging with one another but also espoused a belief that apart and beyond the Howard experience there was an expectation for students to continue to be social agents of change. According to Jayla, at Howard there was an expectation of students to "ask for what you want" and to "seek out opportunities" (J. Capleton, personal communication, August 15, 2005). This example of personal agency is critical at the doctoral level, where there is often little guidance to becoming a scholar and the process of successfully negotiating barriers to success (by way of seeking assistance) can be critical to degree completion (Margolis and Romero, 1998). Furthermore, Jayla's exposure to people of different social classes, ethnicities, and heritages provided her with a rich cultural experience.

MICHAEL (MOREHOUSE COLLEGE)

Michael, who completed his doctoral degree at the age of 26, is a scholar in the field of anthropology of education with a specialization in hip-hop culture and youth identity development. He attended Morehouse College and was motivated to enroll because of its strong academic reputation (in fact, he likened Morehouse to a "Black Harvard") (M. Jones, personal communication, June 8, 2005). Additionally, his brother, a former Morehouse student, influenced him to enroll. Michael offers a unique perspective about the dynamics of the social networks on campus. He indicated that the social environment was intellectually rich with most social gatherings focusing on scholarly development, character building, and encouraging a sense of Black identity. For example, he notes in particular,

> There are things like King's Chapel where every Tuesday a Black speaker would come in . . . outside of the formal curriculum. This ideal of becoming a Black intellectual and being someone who is a legitimate and viable gave us a sense of pride in a Black intellectual traditional setting.
> (M. Jones, personal communication, June 8, 2005)

Michael's reflections on how Morehouse prepared him academically for doctoral study in the Ivy League are as follows:

> [Those] who went to Morehouse were very well positioned when they enter doctoral study in the Ivy League. I think most people who go from Bachelor to Ph.D. programs often have access to certain types of mentorship.
> (M. Jones, personal communication, June 8, 2005)

According to Michael, scholarly development was highly valued by the Morehouse community. It appears that the ideal of scholarship and intellectualism was an integral part of most of the general institutional and communal activities on campus. Like the notions set forth by Allen (1992), Michael affirmed that activities outside of the formal curriculum legitimized and supported not only the academic growth of Morehouse students, but grounded that development in a scholarly and social format. Similarly, his account supports Tinto's (1987, 1993) finding that students who are socially well integrated often attain higher levels of academic achievement. His statements further suggest that students who attend institutions like Morehouse are exposed to academic socialization factors significant to doctoral student development.

JAMES (MOREHOUSE COLLEGE)

In 1994 James received his PhD in education from an Ivy League institution. He received his undergraduate degree from Morehouse College in history. James echoed some of Michael's previous statements concerning the academic rigor promoted in the HBCU environment. He explained how his experience at Morehouse academically prepared him for doctoral study in the Ivy League (J.Williams, personal communication, May 17, 2005). Specifically, he commented on the rigorous studying schedule he imposed on himself while he was engaged in doctoral study, which he attributes to his academic preparation at Morehouse. James stated,

> I would study around the clock; reading every line of every page of my assignments. On the weekends I would stay in the library from 9 am to 5 pm while many other people were doing other things. I felt called to be prepared for class. At Morehouse, it was insulting to a professor to come to class unprepared.
> (J. Williams, personal communication, May 17, 2005)

He states how his undergraduate academic experiences prepared him for graduate-level research activities.

> Morehouse definitely prepared me because I was a history major. We had to do a lot of research projects, especially in our senior year. We had to do a mini-thesis ourselves and all our papers we had to present in front of our peers and we would have to defend our papers. This definitely prepared me for graduate school.
> (J. Williams, personal communication, May 17, 2005)

According to James, for some Morehouse students the strength and life span of the social networks at HBCUs are often developed well beyond a student's separation from their *alma mater*. In fact, during our interview he commented that he would be attending a social function at Morehouse and explained, "I enjoyed my experience at Morehouse. I'm going back. I definitely have fond memories because it helped me to get where I am" (J. Williams, personal communication, May 17, 2005). This exemplifies that James's participation in the social networks at Morehouse were intact well after he completed his doctoral study and supports Allen's (1992) notion that student relationships with faculty members on an HBCU campus can influence the quality of life for its students.

WILMA (HAMPTON UNIVERSITY)

Wilma, a scholar in the field of elementary education, researches multicultural children's literature and how teachers effectively use this literature in the classroom. Of the four interview participants presented, she was the only one married with children, lending an interesting perspective to her responses. She discussed the challenges of balancing family and academics, and credits her undergraduate institution for providing a culturally supportive environment that prepared her for these challenges. Similar to her research interests about student-teacher interactions at the elementary school level, one of her interview responses was sensitive to the peer-peer interaction during her undergraduate academic experience at Hampton:

> In terms of academics, I think it was a good experience for me. Where I had gone previously; my high school experience was in a predominately White environment. So it was good for me academically to be around a lot of African American students

who were also pretty high achievers as well. That was of one of the reasons I wanted to go to Hampton for that and to be more immersed in more of a culturally, at that time what I thought was a more culturally responsive school environment.

(W. Bryant, personal communication, July 18, 2005)

This quote represents some of what Fries-Britt (1998, p. 570) found in her work about high achieving African Americans within "race-specific" communities. She notes that proponents of these communities advocate that they make a tremendous positive impact on the social and academic development of its members.

While Wilma commented on Hampton's race-specific environment, similar to Jayla, she also describes the diversity of Hampton's Black community:

It was primarily African American or students of some type of African descent. Because it is a historically Black college, that is usually the demographic representation. Although, within that population, I would say it was very diverse in terms of the different economic backgrounds students came from and geographic areas.

(W. Bryant, personal communication, July 18, 2005)

According to Fries-Britt (1998), many Blacks shy away from race-specific institutions citing the pressure to develop a sense of belonging in historically White environments. While this point is worth consideration, evidence from these interview responses refutes this claim and shows that race-specific institutions can serve to enhance the peer-to-peer interaction and the psychosocial and academic experiences of Black students. In addition, Black colleges serve to diminish high achiever isolation and consequently support Black undergraduate students who decide to enter doctoral study by providing culturally rich environments that promote comfortable learning experiences.

DISCUSSION AND IMPLICATIONS

Based on the previously mentioned issue of increasing the numbers of minority faculty, it is worth noting that three of the four participants who attended HBCUs are now serving in tenure-track faculty positions. Their experiences demonstrate how the undergraduate preparatory environments of HBCUs provide an inimitable and significant experience for African Americans preparing to enter doctoral study. As indicated in the previous interview responses, this readiness prepares them for the academic and social integration necessary for negotiating barriers to success at the doctoral level work and sets a foundation that guides them for life.

It appears, based on the interview responses that students who enter doctoral study with prior socialization experiences, rich with academic preparation that is rigorous similar to doctoral program activities; social networks that involve intellectual and scholarly development; and participation in environments that are culturally and spiritually sensitive (serving to educate in a holistic context), perform well at the doctoral level within a Ivy League environment. While this research area is still very underdeveloped, this study underscores that the HBCU experience can be a critical factor in the persistence and development of African Americans who matriculate into doctoral programs at Ivy League institutions (Allen, 1992; Thompson, 2005; Wolf-Wendel, 1998; Hood and Freeman, 1995).

In particular, the literature in this study notes the importance of the psychosocial and academic development of these students during their undergraduate experiences at HBCUs (Hughes, 1987; Allen, 1992; Thompson, 2005). The participants' indication of their belief systems and behaviors about this development seems to be in concert with these assertions. Hence, given the slowly growing body of evidence of the HBCU experience as an important factor in African American doctoral student success, more research about this relationship can serve to illuminate the positive characteristics of these institutions in relation to doctoral student preparation. To that end, there needs to be more documentation of HBCUs and how they prepare students for doctoral study and faculty careers. As with this study, documentation that involves qualitative inquiry can offer detailed description of data that often reflects student experience and highlights the impact of HBCUs on student success and doctoral degree completion. Furthermore, interpretation of African American belief systems and behaviors of the HBCU-Ivy League doctoral preparation relationship can be valuable in understanding the nature and context of this connection that improves African American doctoral degree completion.

Elite colleges and universities that struggle with issues of racial inclusion may find it challenging to diversify at all levels of the institution. One way to address this challenge is to develop alliances between HBCUs and Ivy League institutions to jointly address preparation strategies for doctoral students within environments that lack diversification. In addition, there is a need to develop more research concerning the beliefs students have about the HBCU-Ivy League transition process. This could further illuminate how HBCUs have cultivated relationships with students who go on to receive doctoral degrees in elite educational environments.

REFERENCES

Allen, W. R. 1992. The color of success: African-American college student outcomes at predominantly White and historically Black public colleges and universities. *Harvard Educational Review* 62(1): 26–42.

Anderson, E. F., and F. A. Hrabowski. 1977. Graduate school success of Black students from White colleges and Black colleges. *Journal of Higher Education* 48(3): 294–303.

Astin, A. W. 1962. "Productivity" of undergraduate institutions: New analyses show that a college's output of doctors of philosophy depends largely on its input of students. *Science* 136(3511): 129–35.

Blackwell, J. 1987. Mentoring and networking among Blacks. In *In pursuit of equality in higher education*, ed. A. S. Pruitt, 146–62. Dix Hills, NY: General Hall.

Bonner, F. A., and M. Evans. 2004. Can you hear me?: Voices and experiences of African American students in higher education. In *A long way to go: Conversations about race by African American faculty and graduate students*, ed. D. Cleveland. New York: Peter Lang.

Boote, D., and P. Beile. 2005. Scholars before researchers: On the Centrality of the Dissertation Literature Review in Research Preparation. *Educational Researcher* 34(6): 3–15.

Brazziel, M. E., and W. F. Brazziel. 1987. Impact of support for graduate study on program completion of Black doctorate recipients. *Journal of Negro Education* 56(2): 145–51.

Brazziel, W. F. 1983. Baccalaureate college origins of Black doctorate recipients. *Journal of Negro Education* 52(2): 102–9.

Davidson, M. N., and L. Foster-Johnson. 2001. Mentoring in the preparation of graduate researchers of color. *Review of Educational Research* 71(4): 549–74.

Fries-Britt, S. 1998. Moving beyond Black achiever isolation: Experiences of gifted Black collegians. *Journal of Higher Education* 69(5): 556–76.

Garibaldi, A. M. 1997. Four decades of progress . . . and decline: An assessment of African American educational attainment. *Journal of Negro Education* 66(2): 105–20.

Gasman, M., and S. Anderson-Thompkins. 2003. Fund raising from Black college alumni: Successful strategies for supporting alma mater. Washington, DC: Council for Advancement in Education.

Hood, S., and D. Freeman. 1995. Where do students of color earn doctorates in education?: The top 25 colleges and schools in education. *Journal of Negro Education* 64(4): 423–36.

Hughes, M. S. 1987. Black students' participation in higher education. *Journal of College Student Development* 6(28): 532–45.

Kerlin, S. P. 1995. Pursuit of the Ph.D.: Survival of the fittest or is it time for a new approach? *Education Policy Analysis Archives* 3(16): 1–20.

Margolis, E., and M. Romero. 1998. The department is very male, very White, very old and very conservative: The hidden curriculum in graduate sociology departments. *Harvard Educational Review* 68(1): 1–32.

Moses, Y. 1989. Black women in academe: Issues and strategies. Washington, DC: Association of American Colleges.

National Science Foundation, National Institutes of Health, National Endowment for the Humanities, United States Department of Education, United States Department of Agriculture, and the National Aeronautic and Space Administration. 2001. Survey of Earned Doctorates.

———. 2003. Survey of Earned Doctorates.

Palmer, R., and M. Gasman. 2008. "It Takes a Village": Social capital and academic success at Historically Black Colleges and Universities. *Journal of College Student Development* 49(1): 1–19.

Patton, L., and F. Bonner. 2001. Advising the historically Black Greek letter organization (BGLO): A reason for angst or euphoria? *National Association of Student Affairs Professionals Journal* 4(1): 17–30.

Perna, L. 2001. The contribution of historically Black colleges and universities to the preparation of African Americans for faculty careers. *Research in Higher Education* 42(3): 267–94.

Preparing Future Faculty. 2006. Overview of preparing future faculty program. Retrieved January 18, 2006, from http://www.preparing-faculty.org/default.htm#about.

Taylor, E., and J. S. Antony. 2000. Stereotype threat reduction and wise schooling: Towards the successful socialization of African American doctoral students in education. *Journal of Negro Education* 69(3): 184–98.

The Ph.D. Completion Project. 2006. Featured profile–Howard university. Retrieved January 20, 2006, from http://www.phdcompletion.org/features/index.asp.

Thompson, P. F. 2005. African American doctoral degree completers and the factors that influence their success within the Ivy League. Doctoral Dissertation, University of Pennsylvania, 2005.

Tinto, V. 1993. Leaving college: Rethinking the causes and cures of student attrition. 2nd Edition. Chicago: University of Chicago Press.

Willie, C. V., M. K. Grady, and R. O. Hope. 1991. *African Americans and the doctoral experience: Implications for policy.* New York: Teachers College Press.

Wilson, R. 1988. Developing leadership: Blacks in graduate and professional schools. *Journal of Black Studies* 19(2): 163–73.

Wolf-Wendel, L. E. 1998. Models of excellence: The baccalaureate origins of successful European women, African American women, and Latinas. *Journal of Higher Education* 69(2): 141–86.

———. 1994. Vital signs: The Statistics that describe the present and suggest the future of African Americans in higher education. *Journal of Blacks in Higher Education* 3, 33–43.

———. 2001. Two historically Black colleges show greater Ph.D. productivity than do Dartmouth, Emory, Vanderbilt, and the University of Michigan: The performance of these Black colleges is superior to 95 percent of all four-year predominantly White colleges. *Journal of Blacks in Higher Education* 34, 122–6.

———. 2002. The number of Blacks completing doctoral degrees declines for the first time in seven years. *Journal of Blacks in Higher Education* 38, 6–9.

CHAPTER 3

BENNETT AND SPELMAN COLLEGES: CREATING BLACK FEMALE PHDS IN THE SCIENCES

SHANNON GARY

Black women have a double burden to bear in life. First, they must combat the assertion that they are not capable of particular types of work because of their gender. For example, women, and specifically Black women, have been underrepresented in the natural and physical sciences, engineering, and mathematics (Leslie et al., 1998; Hanson, 2004; Salters, 1997). Second, Black women struggle with the prejudice that accompanies being Black in a country that continues to toil with racism. The combination of these influences has often prevented Black women from pursuing education and science-based professions within these fields. As a result, "the culture of science was historically, and is currently, a male culture that is often hostile to women and minorities" (Hanson, 2004, p. 99).

Scholars have conducted a great deal of research on Blacks in higher education (Allen, 1992; Allen and Jewell, 2002; Drewry and Doermann, 2001; Price, 1998; Redd, 1998, 2000). Within this field, some researchers stress the importance and success of historically Black colleges and universities (HBCUs) within the United States, often focusing on the characteristics of Black colleges and how they positively affect the educational outcomes of the students who attend them (Allen, 1992; Redd, 2000; National Science Foundation, Division of Science Resources Statistics, 2006). For example, the National Science Foundation (2006) has shown that HBCUs are sending a disproportionate number of Black students to graduate school, despite the unflattering mainstream perception of these institutions and their students (Watkins, 2001). However, when comparing Black colleges to historically White institutions (HWIs) it is important to understand that the use of traditional quality measures, such as the SAT scores of enrolled students, does not fully illustrate the strengths of HBCUs or their students. Furr and Elling (2002) suggest that traditional measures of academic ability are not reliable predictors of future academic performance for Black students. In fact, Black students often perform beyond what traditional predictive instruments project (Pruitt and Isaac, 1985).

The negative critiques of Black colleges often fail to recognize the opportunity these institutions provide their student bodies. According to Brown et al. (2001), some argue that HBCUs continue to be vestiges of segregation. Similarly, critics state that the open admission policy of many Black colleges lowers the academic quality of these institutions. However, these critiques have been dispelled:

> More than any other set of institutions, HBCUs have discarded the notion that higher education is an advantage open only to the rich or socially prestigious . . . these colleges and universities have established and maintained a tradition of academic excellence.
>
> (Brown et al., 2001, p. 568)

For instance, Spelman College sent a team of six undergraduates to compete in the RoboCup 2005 competition held in Osaka, Japan. Spelman was one of five teams from the United States. It was the only undergraduate institution, as well as the only all women's institution to qualify for the competition (*Black Issues in Higher Education*, 2005). Likewise, HBCUs are responsible for Blacks pursuing education and careers in fields in which minorities are traditionally underrepresented. During an interview with Grayson Walker (2005), Norman Francis, president of Xavier University of Louisiana, stated that since 1927 Xavier University has produced 25% of the Black pharmacist practicing in the United States. As of May 2005 (pre-Hurricane Katrina), 60% of Xavier's undergraduate population was majoring in the natural sciences. Nearly half of those students continue on to graduate school.

To examine why Black colleges have successfully sent Black students to doctoral programs, I will review the history of Spelman and Bennett colleges within the context of the larger Black college literature, looking at their original purpose and why they continue to exist today. It is essential to understand the purposes of historically Black women's colleges and how their histories continue to guide their current mission. I will then consider how these colleges have changed since their inception. More importantly, I will look at the potential benefits of those changes and how they legitimize these institutions.

This chapter will focus on Black women pursuing degrees in math and science.[1] Specifically, I will examine how two institutions, Spelman College and Bennett College for Women, both historically Black women's colleges, have produced an inordinate number of Black women who are PhDs or are currently pursuing their PhD in science or mathematics. Together, these two institutions are responsible for 50% of the Black women pursuing doctoral degrees in the sciences (Gasman, 2005; Frederick D. Patterson Research Institute, 2007). Black women pursuing their PhD in the sciences overwhelmingly receive their baccalaureate degree from specialized institutions such as HBCUs or women's colleges (Hanson, 2004; Solorzano, 1995). These schools are especially prepared to develop students in a society that practices gender and racial inequalities because of their ability to concentrate their efforts toward a specific group.

SPELMAN COLLEGE

The American Baptist Home Mission Society (ABHMS) founded Spelman College in 1881, as the Atlanta Baptist Female Seminary, in the basement of Friendship Baptist Church in Atlanta, Georgia. The seminary was later renamed Spelman Seminary to honor the family of John D. Rockefeller's wife. At the time of Spelman's founding, Rockefeller was one of the largest benefactors of Black colleges. The two founders, Sophia Packard and Harriet Giles, thought it imperative to have an institution to develop Black females into citizens with acceptable social grace and manners befitting a woman of that time, an understanding and appreciation of Christianity, and an ability to teach these qualities to others in their community (Watson and Gregory, 2005). Spelman's founders designed the original curriculum to both educate women for wage earning and provide them with religious education for family life (Brazzell, 1992). As with most Black colleges at the time, Spelman's curriculum focused on elementary education, which was crucial because 90% of the adult Black population in the South was illiterate (Anderson, 1988).

Packard and Giles were both graduates of Northeastern women's colleges; and they modeled Spelman after the preeminent women's institutions of the time. Like its male counterpart, the Atlanta Baptist Seminary (later renamed Morehouse College), Spelman was dedicated to creating leadership within the Black community (Watson and Gregory, 2005). Spelman's founders viewed the alumnae as future leaders of the Black race, and as such, students were obligated to disseminate their recently acquired religious ideas to the Black masses. According to Brazzell (1992), leaders of the ABHMS were the first to develop the idea of a talented tenth, most widely associated with W. E. B. Du Bois. The ABHMS believed it was the organization's responsibility to educate a talented minority of the Black race to lead other Blacks to equality within American society, "one of the main tenets of Baptist missionary doctrine . . . collegiate education as the process by which a cadre of Black leaders would be developed to uplift the masses of Black people" (Brazzell, 1992, p. 36).

Today, Spelman is one of the most revered Black institutions in the country. The institution is nationally ranked among liberal arts colleges and boasts one of the highest SAT scores and lowest acceptance rates among Black colleges. Spelman's SAT scores range between 980 at the 25th percentile and 1160 at the 75th percentile and the institution has a 37% acceptance rate, making it one of the top 75 liberal arts colleges in the United States (*U.S. News and World Report*, 2007). The selective nature of Spelman began during the early 1900s under the administration of Lucy Tapley. During her administration, Spelman began its use of intellectual assessments, for both prospective and current students, to properly place students in appropriate educational tracks (Watson and Gregory, 2005). This began Spelman's recognition as a "true college." For example, Tapley built a science building to provide appropriate instructional and lab space for students in the sciences. The college also ended the school's reliance on Morehouse College's science facilities. By 1930, Spelman was a fully established college, no longer offering elementary or high school education, collaborating with Morehouse College and Atlanta

University to create the Atlanta University Center. Since then the institution has continued to train women in all realms of study. Spelman College became and still remains a beacon of success, attracting students to its campus from all over the United States.

BENNETT COLLEGE FOR WOMEN

Bennett College has not received the same attention as Spelman College, resulting in a smaller amount of scholarly research on the institution. Unlike Spelman College, the Freedman's Aid Society of the Methodist Episcopal Church originally founded Bennett College in 1873 as a coeducational institution in Greensboro, North Carolina, in the basement of the Warnersville Methodist Episcopal Church. The school was later named for Lyman Bennett, a New York businessman who donated a significant sum of money, which resulted in the school moving to its first permanent home. By 1886, Bennett served as an industrial workplace for women and had become a single-sex institution because of World War I and the war's influence on societal changes that took place for both Blacks and women in the South. Bennett's initial curriculum focused on preparing women for marriage and home life (Bell-Scott, 1984). However, the college also developed leadership within its students. "They . . . were clearly provided with the tools to move beyond the home and pursue leadership and professional roles in the community after graduation" (Liberti, 2004). The college's record of influential alumnae is as impressive as other exceptional Black colleges. Notable Bennett alumnae include Glenora M. Putnam, the first Black woman to serve as president of the National Young Women's Christian Association (YWCA), and Hatie Carwell, a research scientist and expert in the study of radiation.

Under the 30-year leadership of David Dallas Jones, Bennett expanded to a 42-acre campus with 33 buildings and a $1.5 million endowment. During his tenure, the college made immense progress and discontinued the high school curriculum once offered. Willa Player became the president after Jones's death in 1956. Player was the first female president of Bennett College and one of the first Black female presidents of any four-year liberal arts institution in the country. During Player's administration, the Southern Association of Colleges and Schools (SACS) granted Bennett accreditation. This move was an indication of Bennett's significance as an educational institution for Black women. Bennett was one of 15 four-year Black colleges at the time to earn accreditation from SACS.

Bennett is currently an institution that offers open admission to all qualified students, and typically draws the majority of its student body from the Southeastern United States. The college experienced difficult times before Johnnetta Cole took the helm of leadership in 2002. Under her leadership Bennett has been removed from SACS probation, the deficit has been erased, and there are new academic programs and increases in enrollment, alumnae giving, and foundation support (Owens, 2005). Bennett demonstrates an institution's ability to transform underdeveloped students into competitive applicants to graduate school. Whereas Spelman focuses on educating a highly selective group of young Black women, Bennett's open admissions policy is more inclusive to those seeking to obtain a baccalaureate degree, regardless of academic adeptness (Owen, 2005), which makes the institution's record for developing Black female PhDs in the sciences all the more compelling.

HBCU CHARACTERISTICS AND ENVIRONMENT

The environment of HBCUs specifically focuses on the academic, social, and psychological development of Black students (Culotta, 1992; Wenglinsky, 1996, 1997; Wagener and Nettles, 1998; Redd, 2000; Allen, 1992; Davis, 1985). These aspects play an important role in the performance of Black students (Allen, 1992; Tinto, 1987), as Black colleges are uniquely equipped to serve and educate this population (Wenglinsky, 1996, 1997; Anderson and Hrabowski, 1977). HWIs are not able to fully duplicate this experience for Black students on their campuses. Based on their institutional histories, curriculum and student development are Euro-centric in design at HWIs (Johnson, 2003). The lack of minorities in the upper echelon of leadership, among the faculty and the Board of Trustees, inherently prevents the full consideration of issues faced by many Black students. Instead, there is an emphasis on the need for Blacks and other minority students to assimilate into the predominately White culture of the campus (Gasman, 2005; Johnson, 2003). Most HWIs have not had the history of specifically serving a persecuted group (Allen and Jewell, 2002), nor do they have the experience of addressing the needs of Black students in meaningful and salient ways since their creation (Allen, 1992). The mission of Black Colleges has been to educate a group of people who were viewed in the past, and arguably today, as second-class citizens of the United States (Allen, 1992; Allen and Jewell, 2002; Wagener and Nettles, 1998; Wenglinsky, 1996). Black colleges were developed to promote racial uplift and serve as institutions for all citizens, regardless of their racial or social backgrounds (Allen, 1992; Allen and Jewell, 2002; Brazzell, 1992; Brown et al., 2001; Drewry and Doermann, 2001). The experience of serving Black students enables HBCUs to be cognizant of the challenges this perception creates within students. This insight allows HBCUs to purposefully address these issues and prevent negative effects such as low self-esteem and low expectations among students, both personally and academically.

Black colleges have continually educated a disproportionate number of Blacks in the United States and have produced a large number of Black graduate students. While there are 103 HBCUs in the United States, these institutions enroll approximately 16% of the Black undergraduate student population (Gasman, 2005; Redd, 2000; Wenglinsky, 1996, 1997). Furthermore, according to Solorzano (1995), 50 colleges and universities in the United States produce 47% of Black females who eventually receive their PhD. Of those 50 institutions, 33 are HBCUs. Likewise, 53 institutions produce 38% of Black males who eventually receive their PhD. Of those 53 institutions, 35 are HBCUs. Black colleges have achieved these results specifically because they have not abandoned their mission of providing Black students an education in a nurturing environment and an opportunity to be trained professionally for both academic and labor pursuits (Wenglinsky, 1996, 1997). Black colleges have indoctrinated this philosophy through all aspects of their campus communities (Wagener and Nettles, 1998). Being specialized institutions allows Bennett and Spelman colleges to specifically target and address issues of concern for Black women. A senior at Bennett states, "It's a place to develop as an African-American woman" (Owens, 2005, p. 158).

Allen (1992) ascertained the factors that most influenced Black students to persist at the undergraduate level: academic achievement, social integration, and academic aspirations. Allen's study demonstrated that social interaction is important to the academic performance of Black students and that academic performance is fundamental to a Black student's decision to attend graduate school. Allen further states that the perception of the campus environment, including interaction with the student body, faculty, and administration, largely shapes the student experience and outcomes. According to his study, many of the Black students who attended HWIs described feelings of alienation, hostility, and lack of social integration (Allen, 1992), whereas Black students who attended HBCUs described opposite feelings in a study by Donna Owens (2005). These students described feelings of engagement, support, acceptance, and encouragement. Bennett College senior, Carah Herring, stated, "The academic support and learning are important, but it's not 100 percent about school work. Bennett is educating women holistically" (Owens, 2005, p. 158). Likewise, at Spelman College, a Chemistry professor explained how most faculty members essentially have an open door policy with their students, supporting the multiple dimensions of these students (A. Thompson, personal communication, March 26, 2006).

Allen's (1992) findings cannot be generalized because of their limited scope. First, there was a low response rate to his survey. Approximately, 30% of the sample completed the survey. This may skew the results in a particular direction based on why students completed the survey. Lack of response from a majority of the sample may also indicate that only those with extremely negative or positive experiences felt inclined to express their feelings. Second, the study lacks a geographic control. The majority of students who responded to the survey attended a public institution in their home state. Most public HBCUs are located in the South, whereas public HWIs are in all regions of the country. Prejudices and biases toward minorities can be vastly different within different regions of the country. Regardless of the limitations in Allen's (1992) study, many other scholars have found similar reactions and descriptions by Black students when comparing experiences at Black colleges and HWIs (Feagin et al., 1996; Culotta, 1992; Wenglinsky, 1996, 1997; Redd, 2000; Davis, 1985).

Both Bennett and Spelman provide positive environments for their students. For example, at Spelman College it is the duty of upperclassmen to assist underclassmen as they navigate through their college experience (Watson and Gregory, 2005). According to Professor Albert Thompson of Spelman College, peer-to-peer academic support is encouraged and supported by the institution (A. Thompson, personal communication, March 26, 2006). Because of the institutional climate at Black colleges, students attending them believe they are able to achieve high goals. A graduate student in physics at North Carolina State University stated, "I've seen how people come out of those black schools. They exude the message, 'I'm capable'" (Culotta, 1992, p. 1218). The fostering of this attitude allows Black colleges, and Bennett and Spelman in particular, to develop a disproportionate number of female PhDs in math and science today.

FOCUS ON TEACHING

Like HWIs, the majority of HBCUs are not Research I universities and do not have a research focus (Solorzano, 1995). This affords Black colleges, on the whole, the opportunity to concentrate on teaching their students. Of the top ten institutions producing Black women PhDs, all are HBCUs. One is a Research I institution, and the top two are Spelman and Bennett colleges (Wolf-Wendel, 1998). According to Wolf-Wendel (1998, 2000), the race and gender composition of an institution plays a major role in encouraging women to pursue a PhD. For instance, HBCUs are able to focus their time and resources to serve the needs of one population. Of note, these findings make the assumption that women and Blacks are a monolithic group. Although this is a false supposition, there are particular issues that affect the group universally, such as sexism and racism, especially in the sciences and mathematics, regardless of individual characteristics (Hanson, 2004). The ability to focus on the needs of Black students is a strength of Black colleges. "Fans of HBCUs say the numbers reflect what they do best: teach undergraduates" (Culotta, 1992, p. 1216). Less emphasis on scholarship and publishing allows faculty at HBCUs to give a substantial amount of attention to their students (Solorzano, 1995). However, lack of research should not be mistaken for lack of intellectualism.

The faculty members at Black colleges are unique in several ways. They function simultaneously as teacher, advisor, mentor, and scholar (Wolf-Wendel, 1998). According to the Spelman Mathematics professor Sylvia Bozeman, faculty members at Spelman started with programs such as the summer support program for pre-freshman, and from there faculty continued their dedication to students (S. Bozeman, personal communication, March 26, 2006). Faculty commitment to extended hours in assisting, counseling, and academic advising is also part of student success at Spelman (A. Thompson, personal communication, March 26, 2006). At Bennett, faculty and administrators take part in intrusive advising. This form of advising requires advisors to proactively provide guidance (Wolf-Wendel, 1998). A Bennett faculty member states, "It is not in your job description . . . your intuit that is what is needed" (Wolf-Wendel, 1998).

Successful faculty members are clearly committed to the success of the students their institution serves (Wagener and Nettles, 1998). The expectation at Bennett and Spelman is to work with students in numerous ways outside of the classroom, to serve as role models for the women they teach, to be proactive with advising, and to identify problems at an early stage (Wolf-Wendel, 1998). Early recognition of academic problems allows HBCUs to identify deficiencies with students' academic performance and to develop interventions before the problem results in dismissal (Wagener and Nettles, 1998). This sense of concern and priority continues to differentiate HBCUs from HWIs. The environment at Black colleges transforms students who often do not fit the model of academic excellence into graduates who pursue doctoral degrees in science and mathematics.

The work of these professors is essential for developing the students who wish to continue their education at the graduate level. For example, at Spelman, encouragement to pursue graduate school begins in the first year of matriculation. NASA-funded grants and the early exposure of students to research in the laboratory, as well as in the classroom, are two reasons that so many students are choosing to pursue graduate studies at these two colleges (A. Thompson, personal communication, March 26, 2006). Originally the focus on teaching at HBCUs was due to the lack of resources; therefore, excellence in teaching was a requirement in lieu of research (Davis, 1985). While for many HBCUs this situation still exists today (Solorzano, 1995), numerous Black colleges have been able to attract and retain respected and well-known scholars who are also excellent teachers (Davis, 1985).

PREPARING STUDENTS FOR GRADUATE SCHOOL

HBCUs have prepared students for graduate school for many years (Stephens, 1994; *Journal of Blacks in Higher Education*, 1994; Ehrenberg, 1996; Culotta, 1992; Brazziel, 1983; Solorzano, 1995; Allen and Jewell, 2002). According to National Science Foundation (2006), from 1995 to 2004, 46% of the Black women who earned a doctoral degree in a science, technology, engineering, and mathematics (STEM) field received their baccalaureate degree from a Black college. This is remarkable for a set of colleges and universities that represent only 3% of the total number of higher education institutions in the United States. Black colleges like Bennett and Spelman have successfully produced a large number of undergraduate science majors, and thereby have played a pivotal role in increasing the number of Blacks in these fields.

BLACK WOMEN PURSUING THE SCIENCES

The number of women obtaining PhDs in mathematics and the sciences has steadily increased. By 1995, 31% of PhDs in the sciences were awarded to women (Salters, 1997). Early familiarization with the sciences and mathematics has been shown to increase the probability of pursuing a degree in these fields. For example, parents employed as a scientist or a mathematician increases the likelihood of pursuing careers in those fields (Leslie et al., 1998). Given the underrepresentation of racial and ethnic minorities who pursue degrees in the sciences, as well as the historical limitations of access to higher education, it is highly improbable that Black students possess this lifelong exposure. This makes it imperative for Black women to observe other Black women engaging and excelling in these professions. Bennett and Spelman colleges allow this to occur. For example, at Spelman students are exposed to a large number of faculty role models, especially African American females. The faculty is student-centered and nearly all the research proposals submitted request the support of undergraduate students (A. Thompson, personal communication, March 26, 2006).

Males and females do not differ early in life regarding interest in science and math; however, during adolescence female's self-esteem lowers (Leslie et al., 1998). This lowered self-image can adversely affect women's pursuit of math and science. Interestingly, this drop in self-esteem does not occur with Black women (Leslie et al., 1998). Leslie et al. (1998) indicate that this may imply a difference in the socialization of Black females compared to females of other races. This difference can offer important insight into understanding the gender differences that affect pursuit of science degrees by Black women. Of note, 36% of Black scientists (earned PhDs) are female compared to White females, who represent 22% of White scientists (Hanson, 2004).

BENEFITS OF HISTORICALLY BLACK WOMEN'S COLLEGES

Self-image, self-confidence, and self-efficacy are fostered and strengthened at Bennett and Spelman colleges. Both of these institutions have been able to limit the adverse affects of female development by surrounding their students with positive and encouraging Black women. At Spelman, faculty members serve as effective role models. There is a strong emphasis on undergraduate research and mentoring programs to connect students with faculty (A. Thompson, personal communication, March 26, 2006). The clear mission and traditions of "sisterhood" and "kinship" at Black women's colleges foster and promote persistence in these fields. At Spelman, peer-to-peer academic support is encouraged by the college (A. Thompson, personal communication, March 26, 2006). Faculty members at these institutions encourage the students they instruct to learn from them as well as their peers. This allows for peers to support one another, instead of discouraging each other.

Bennett and Spelman challenge the stereotype of females performing poorly in science and math fields (Hanson, 2004; Wolf-Wendel, 2000). In her 1994 book *Conversations: Straight Talk with America's Sister President*, Johnnetta Cole asserted that "at Spelman, there is no assumption that Black folks don't like math and women cannot do science . . . On the contrary, the assumption is that Spelman women will excel in all of their studies" (p. 151). I would suggest she has carried this same philosophy to Bennett College.

Women at these institutions are prohibited from using racism and sexism to explain why they are incapable of achieving in a particular field of study (Wolf-Wendel, 2000). When students doubt their ability, they merely need to observe the individuals at their college to realize what they are capable of achieving. These women are keenly aware of what Blacks generally, and Black women specifically, have accomplished in the fields of math and science (Watson and Gregory, 2005). The success of current students and alumnae offer motivation to aspiring graduates (Wolf-Wendel, 2000). A clear history of what alumnae and other Black women have accomplished in the sciences allows for students to appreciate what they can achieve.

According to Hanson (2004), a major limitation of most studies that examine women pursuing the sciences is the focus on women as a single entity. There is very little research on the ethnic and racial differences of women pursing professions in the sciences and math (Hanson, 2004). The divergent experiences and psychosocial development of Black women and White women play a large role in their choice of and persistence in math and science professions (Leslie et al., 1998). Racial and cultural expectations provide differing views of responsibility to family and gender roles (Brazzell, 1992). Black women's gender roles have been blurred since slavery (Brazzell, 1992), unlike the role of White women within their culture. These findings explain Hanson's (2004) conclusion that Black women do not see careers as conflicting with family; instead, they see career as a part of motherhood. This often explains the higher career expectations Black women have for themselves and their families.

Black women are more motivated when they have high racial identity because they are more likely to associate scholarly achievement with racial identity (Cokley, 2003). Bennett and Spelman offer the opportunity for this identity to be cultivated. For example, at Bennett, freshmen have the opportunity to participate in The Summer Academy. According to Bennett's website, the purpose of the academy is to provide students with information about campus resources and support services and to participate in "cultural identity" activities. Additionally, the academy aims to develop "lifelong Bennett sisterhood" in a supportive and inclusive environment. Likewise, Watson and Gregory (2005) maintain, "The term *sister* at Spelman denotes a common term of endearment for students, faculty, administrators and alumnae. The value of sisterhood and community at Spelman has been encouraged through a nurturing environment" (Watson and Gregory, 2005, p. 144). In contrast, the anxiety of Black females is heightened at both coeducational HBCUs and HWIs (Cokley, 2003). As a result, the productivity ratio of producing women in the sciences at Bennett and Spelman is 47 times higher than historically White coeducational institutions (Wolf-Wendel, 1998).

Bennett and Spelman colleges are producing a disproportionate number of Black women obtaining graduate degrees in the sciences (Gasman, 2005). They have used their strengths to develop young Black women into competitive candidates for STEM doctoral degrees by providing their students with opportunities to progress beyond both social and economic limitations that exist for a large portion of Black college student bodies.

Wolf-Wendel (1998, 2000) states that the racial and gender composition of an institution has a critical influence on the decision of Black women to pursue doctoral graduate study. Given the single mission focus of these two colleges, they are able to specifically address the academic, social, and psychological development of their students. Women at these institutions are able to depend on faculty members as well as administrators to help guide them as mentors, role models, and teachers. Likewise, the culture at both colleges has fostered the formation of a bond between the women in attendance that has allowed students to develop a sense of reliance on one another. These women do not view their "sisters" as rivals, but instead they regard each other as equals. As such, they have learned to support one another. Peer support is significant to the academic success of Black females (Watson and Gregory, 2005). These campuses have created cultures of engagement, support, and encouragement (Owens, 2005) from all members of the community. This type of environment has proven to be crucial in developing the confidence of women, but even more so for Black women.

Continued research needs to be conducted to identify factors that contribute to the persistence of Black women in the sciences. Likewise, thorough examination of the institutions that are succeeding in this endeavor must persist so their practices can be replicated. These changes will not occur immediately; however, serious effort must be exercised by all institutions, HBCUs, and HWIs to ensure that Black women continue to pursue degrees in the sciences and mathematics. Black women are an untapped resource.

NOTE

1. The term "science" includes all physical sciences, life sciences, and engineering.

REFERENCES

Allen, W. R. 1992. The color of success: African-American college student outcomes at predominately White and historically Black public colleges and universities. *Harvard Educational Review* 62(1): 26–44.

Allen, W. R., and J. O. Jewell. 2002. A backward glance forward: Past, present, future perspectives on historically Black colleges and universities. *Review of Higher Education* 25(3): 241–61.

Anderson, E. F., and F. A. Hrabowski. 1977. Graduate school success of Black students from White colleges and Black colleges. *Journal of Higher Education* 48(3): 294–303.

Bell-Scott, P. 1984. Black women's higher education: Our legacy. *SAGE: A Scholarly Journal on Black Women* 1(1): 8–11.

Brazzell, J. C. 1992. Bricks without straw: Missionary sponsored Black higher education in the post-emancipation era. *Journal of Higher Education* 63(1): 26–47.

Brazziel, W. F. 1983. Baccalaureate college of origin of Black doctorates. *Journal of Negro Education* 52(2): 102–9.

Brown II, M. C., S. Donahoo, and R. D. Bertrand. 2001. The Black college and the quest for educational opportunity. *Urban Education* 36(5): 553–71.

Cokley, K. O. 2003. What do we know about the motivation of African American students? Challenging the "anti-intellectual myth." *Harvard Education Review* 73(4): 529–54.

Cole, J. 1994. *Conversations: Straight talk with America's sister president.* New York: Anchors Books.

Culotta, E. 1992. Black colleges cultivate scientist. *Science* 258(5085): 1216–8.

Davis, A. L. 1985. The role of Black colleges and Black law school in the training of Black lawyers and judges: 1960–1980. *Journal of Negro Education* 70(1/2): 24–34.

Drewry, H. N., and H. Doermann. 2001. *Stand and prosper: Private Black colleges and their students.* New Jersey: Princeton University Press.

Ehrenberg, R. G. 1996. Are Black colleges producing today's African-American lawyers? *Journal of Blacks in Higher Education* 14: 11719.

Feagin, J. R., H. Vera, and N. Imani. 1996. *The agony of education: Black students at White colleges and universities.* New York: Routledge.

Frederick D. Patterson Research Institute (2007). http://www.patterson-uncf.org/.

Furr, S. R., and T. W. Elling. 2002. African-American students in a predominately White university: Factors associated with retention. *College Student Journal* 36(2): 188–203.

Gasman, M. 2005. What can Penn learn from our nation's historically Black colleges and universities (HBCUs)? *University of Pennsylvania Almanac* 52(11): 8.

Guy-Sheftall, B. 1982. Black women and higher education: Spelman and Bennett colleges revisted. *Journal of Negro Education* 51(3): 278–87.

Hanson, S. 2004. African American women in science: Experiences from high school through the post-secondary years and beyond. *National Women's Studies Association* 16(1): 96–115.

Hilton, T. L., and V. E. Lee. 1998. Student interest and persistence. *Journal of Higher Education* 59(5): 510–26.

Johnson, V. D. 2003. A comparison of European and African-based psychologies and their implication for African-American college student development. *Journal of Black Studies* 33(6): 817–29.

Leslie, L., G. T. McClure, and R. L. Oaxaca. 1998. Women and minorities in science and engineering: A life sequence analysis. *Journal of Higher Education* 69(3): 239–76.

Liberti, R. 2004. "We were ladies, we just played like boys": African-American woman-hood and competitive basketball at Bennett College, 1928–1942. In *Sports and the color line: Black athletes and race relations in twentieth century America,* ed. P. B. Miller and D. K. Wiggins. New York: Routledge.

National Science Foundation, Division of Science Resources Statistics. 2006. *S&E Degrees by Race/Ethnicities of Recipients: 1995–2004.* NSF 07–308. Susan T. Hill and Maurya M. Green, project officers, Arlington, VA.

Owen, D. M. 2005. Daughters of the dream. *Essence* 35(10): 154–8.

Price, G. 1998. Black colleges and universities: The road to philistia? *Negro Educational Review* 49(1–2): 21.

Pruitt, A., and P. Isaac. 1985. Discrimination in recruitment, admission and retention of minority graduate students. *Journal of Negro Education* 54(4): 526–36.

Redd, K. E. 1998. Historically Black college and universities: Making a comeback. *New Directions in Higher Education* 102: 33–43.

———. 2000. HBCU graduates: Employment earnings and success after college. *New Agenda Series* 2(4): 1–22.

Salters, R. E. 1997. Pursuing the Ph.D. in the sciences and engineering: Trends and observations. *New Directions in Higher Education* 99: 91–7.

Solorzano, D. G. 1995. The doctorate production and baccalaureate origins of African-Americans in the sciences and engineering. *Journal of Negro Education* 64(1): 15–32.

Stephens, L. F. 1994. *Motivations for enrollment in graduate and professional school among African-American students in HBCUs.* South Carolina: Houston Center for the Study of the Black Experience Affecting Higher Education.

Tinto, V. 1987. *Leaving college: Rethinking the causes and cures of student attrition.* Chicago: University of Chicago Press.

Wagener, U., and M. T. Nettles. 1998. It takes a community to educate students. *Change* 30(2): 18–25.

Walker, G. 2005. Corporate spotlight–Xavier University. *American Executive.* Available at http://www.americanexecutive.com/content/view/4574/79/.

Watkins, W. 2001. *The White architects of Black education.* New York: Teachers College Press.

Watson, Y. L. and S. T. Gregory. 2005. *Daring to educate: The legacy of the early Spelman college presidents.* Sterling, VA: Stylus.

Wenglinsky, H. H. 1996. The educational justification of historically Black colleges and universities:

A policy response to the U.S. supreme court. *Educational Evaluation and Policy Analysis* 18(1): 91–103.

———. 1997. *Students at historically Black colleges and universities: Their aspirations and accomplishments.* Princeton, NJ: Educational Testing Services.

Wolf-Wendel, L. 1998. Models of excellence: The baccalaureate origins of successful European American women, African American women and Latinas. *Review of Higher Education* 69(2): 141–86.

———. 2000. Women-friendly campuses: What five institutions are doing right. *Review of Higher Education* 23(3): 319–45.

1994. Hampton University: The new breeding ground for the Black physicist. *Journal of Blacks in Higher Education* 14: 117–9.

1997. Black colleges holding their own in preparing students for doctorates in the sciences. *Journal of Blacks in Higher Education* 16: 58–9.

2002. Two historically Black colleges show greater Ph.D. productivity than do Dartmouth, Emory, Vanderbilt, and the University of Michigan: The performance of these Black colleges is superior to 95 percent of all four-year predominantly White colleges. *Journal of Blacks in Higher Education* 34: 117–9.

2005. Spelman students shatter myths about women as leaders in science. *Black Issues in Higher Education* 22(12): 12.

2007. *US News and World Report.* 2008 Edition, "America's Best Colleges"

CHAPTER 4

SOCIAL JUSTICE, VISIONARY, AND CAREER PROJECT: THE DISCOURSES OF BLACK WOMEN LEADERS AT BLACK COLLEGES

GAETANE JEAN-MARIE

Throughout the history of African Americans, Black women have played an integral role in the social movement for equal educational opportunity for people of color. In their personal quest for an education, they were susceptible to interlocking systems of sexism and racism that permeated public and higher education. For many Black women, their coming of age was inextricably linked to the larger changing consciousness of African Americans who challenged the existing social order in new ways (Ladson-Billings, 1997; Robnett, 1997). The purpose of this chapter is to examine three projects—social justice, visionary, and career—of 12 Black women leaders in historically Black colleges and universities (HBCUs) in one southeastern state and how they connect their lived experiences and positions as educational leaders to the mission of developing the educational and social functions/capital of the Black community.

The distinctions between the three projects portray the multiple and competing ways Black women interpret their work as leaders in HBCUs. A *career project rooted in individual achievement* places emphasis on the individual leader who practices top-down and authoritarian way of leading. The institution moves forward through the actions taken by the leader. A *visionary project rooted in economic success* is based on a corporate mindset operating within educational entities and students are viewed as consumers whose education is the product being sold. For *a social justice project rooted in community*, the participants interpret their leadership in the context of a spiritual or religious realm, and through shared community commitment. Students are regarded as children of the community who are to be nurtured, groomed, and "prayed" for as leaders of the next generation. These women's reflections represent some of the complexities of Black identities and institutions in the twenty-first century. Collins (1998) asserts that dialogues among Black women that are attentive to both heterogeneity among African American women and shared concerns arise from a common social location such as HBCUs. Through this research, the participants had a space to dialogue about their work in HBCUs, which created a collective discourse of Black women leaders.

Introduced are women whose elementary and secondary educational foundations were formed at the cusp of segregation in American schools (i.e., fewer resources, lack of funding, limited teachers and classrooms). Coming of age was inextricably linked to the larger changing consciousness of these African American women who challenged the existing social order in new ways (Ladson-Billings, 1997; Robnett, 1997). Consistent with previous research (Morris, 1994), the women leaders in this study came from a tradition of protest that has been transmitted across generations by older relatives, Black educational institutions, churches, and protest organizations (Jean-Marie, 2005). Like these female leaders, the perspectives of African American women who were born and raised during the pre– and post–Civil Rights Movement are uniquely shaped (Jean-Marie, 2006; Loder, 2005; Robnett, 1997) by their experiences.

THEORETICAL FRAMEWORK

LEADERSHIP, SOCIALIZATION, AND GENDER INFLUENCE

Literature on leadership and gender often focuses on the influence of the latter to emphasize certain dimensions of leadership over others (Pounder and Coleman, 2002). Leadership styles of women are described in general terms as interpersonal-oriented, charismatic, nurturing, and democratic (Eagly et al., 1992; Freeman and Varey, 1997) and related to gender because of stereotypes of women as being sensitive, warm, tactful, and expressive (Olsson and Walker, 2003; Van Engen et al., 2001). The familiar belief is that women need to be trained up to the level of men, arguing training for equal opportunities, rather than value what they bring to organizations (Cubillo and Brown, 2003). Fortunately, the "baggage" that women leaders bring to educational organizations is gaining more visibility and attracting more attention as the female presence in leadership positions increases. The baggage is the result of several basic influential factors that shape behaviors of women throughout their personal and professional lives: (a) socialization, (b) organizational culture, and (c) gender influence and power.

SOCIALIZATION

A major component of any leadership development process involves socialization whereby attention is drawn to the leader and the context simultaneously. Brim (as cited in Feldman, 1989, p. 3) defines socialization as "the manner in which an individual learns the behavior appropriate to his position in a group through interaction with others who hold normative beliefs about what his role should be and who reward or punish him for correct or incorrect actions." Bennis (1985) asserts that within an organization the socialization includes all the people in it and their relationships to each other and to the outside world. Hence, the behavior of one member can have an impact, either directly or indirectly, on the behavior of others.

Socialization of Women

Because of how the socialization process unfolds, women have developed values and beliefs that translate into specific behaviors arising in their leadership styles. Research has indicated that women are socialized to show their emotions, feelings, compassion, patience, and intuition; to help and care for others (Bass and Avolio, 1994; DeMatteo, 1994; Greenleaf, 1996; Pounder and Coleman, 2002); to be listeners (Brunner, 1998); to judge outcomes based on their impact on relationships (Klenke, 2003; Oakley, 2000); and to lead complex settings in continuous change (Caprioli and Boyer, 2001).

The ways women approach the job of educational leadership are related to the models of leadership they encounter in their careers and to the goals they hope to achieve through their positions as leaders (Normore, 2006; Young and McLeod, 2001). What influences women's leadership styles in education is the *degree of identification* with a leadership model, whether to be adopted or discarded. Research indicates that there are mediating influences on leaders' socialization such as work setting, relationships with peers, superiors, organization policies and procedures, formal training, outcomes, and organizational culture (Normore, 2006; Rutherford, 2001; Schein, 1992). This latter influence incorporates the image of the role of the leader, skills, norms and values, and communication networks (Normore, 2006). As a consequence, it is the organizational culture that highly influences the leadership styles that are predominant and accepted in a particular organization.

ORGANIZATIONAL CULTURE

Culture is defined as the essence of traditional ideas and their attached values to thinking, feeling, and reacting to various stimuli (Kluckhohn, 1951). Organizational culture refers to the set of assumptions, beliefs, values, and norms that are shared by members of an organization and is influenced by its past, environment, and industry (Rutherford, 2001). Organizational culture also applies to communication, codes of behavior, processes, and policies (Still, 1994).

Female educational leaders focus on their primary responsibility, which is the care of children and their academic success (Normore, 2006). Building relationships with others to achieve common goals is a recurrent topic of women in leadership positions in education. Women value close relationships with students, staff, colleagues, parents, and community members as key in school leadership (Furman and Starratt, 2002; Williamson and Hudson, 2001). In schools headed by women, relationships develop constantly through spending time with people, communicating, caring about individual differences, showing concern for teachers and marginal students, and dedicating more energy to motivate others (Grogan, 1994; Williamson and Hudson, 2001). Important for women educational leaders is also communication to keep everybody informed and to reach others (Gronn, 2003; Shuttleworth, 2003; Wesson and Grady, 1993).

Women and Organizational Culture

Women continue "paving the way" through the different organizational cultures in search of leadership styles that are more authentic and less accommodative. Like any new trend in traditional settings, it takes years to develop styles until these styles are understood and accepted. Meanwhile, women face several barriers that prevent them from being considered leaders or even leadership candidates (Still, 1994). Obstacles with this origin have been often described as "the glass ceiling"— as a metaphor of an invisible top that halts women in moving up the career ladder at a certain point (Oakley, 2000). Nevertheless, according to Evans (2001), women are more than ready and are *free* from feminine parameters of leadership to compare to in their trial-and-error quest to develop their own styles. Developing a particular leadership style has become especially prevalent for women in educational leadership roles. For the African American women participants in this study, their leadership approaches are influenced by their experiences during the Civil Rights Movement and their involvement in that struggle to challenge the status quo.

RACE AND GENDER: THE BLACK FEMALE EXPERIENCE

Sexism in institutionalized systems, accompanied by disenchantment with the White-dominated feminist movement and Black male scholars' exclusive concern with racial issues (Schiller, 2000) during the Civil Rights Movement, heightened Black women's interest in liberation. Finding no place in the existing movement and wanting to respond to the racism of White feminists and the sexism of Black men, Black women formed separate "Black feminist" groups (hooks, 1981). They sought to create new knowledge about African American women to "formulate and rearticulate the distinctive, self-defined standpoint of African-American women" (Collins, 1996, p. 225). They also sought to change the one-dimensional perspectives on women's reality.

Through these efforts, African American women positioned themselves to engage in critical analysis by articulating their "voices to express a collective, self-defined Black woman's standpoint" (Collins, 2000, p. 99). Voice defines who they are, interprets what their experiences are, and analyzes their coping mechanisms for survival (Jean-Marie, 2003, 2004). Consequently, African American women develop a double consciousness (Collins, 2000) that empowers them to move in and out of diverse spaces. As more African American women continue to make inroads into professions and occupations previously dominated by Euro-American women (Mullings, 1997), they will likely have an impact on the representation of African American women of all echelons.

Developing an Afrocentric Epistemology

To understand how Black women's consciousness evolved, an understanding of Afrocentric epistemology (Collins, 2000) is vital. Black women's Afrocentric epistemology is the significance and richness of their African roots that inform what they believe to be true about themselves and their experiences. In a society that often devalues heritage, Black women draw from common experiences that historically connect them to the fundamental elements of an Afrocentric standpoint (Jean-Marie, 2004). Because Blacks share a common experience of oppression resulting from colonialism, slavery, apartheid, imperialism, and other systems of racial domination (Collins, 1996), these shared material conditions cultivate Afrocentric values within Black communities throughout the world. As a result, they seep into the family structure, religious institutions, and community. The collective history of people of African descent from Africa, Caribbean, South and North America constitutes an Afrocentric consciousness that permeates through the framework of a distinctive Afrocentric epistemology.

Similarly, women share a history of patriarchal oppression. The persistence of sexism contributes to the exploitation of women (Davis, 1989; Dorn et al., 1997–98). Furthermore, the degree of exploitation is related to social class, race, religion, sexual orientation, and ethnicity (Collins, 1996). Women have a body of knowledge that corresponds with feminist consciousness and epistemology. Because African American women have access to both the Afrocentric and the feminist standpoints, an alternative epistemology is used to rearticulate Black women's standpoint that reflects elements of both traditions (Collins, 2000). Black women's epistemology represents a specialized knowledge that provides opportunities to express Black feminist concerns:

> The experiences of African-American women scholars illustrate how individuals who wish to rearticulate a Black women's standpoint through black feminist thought can be suppressed by prevailing knowledge validation processes.
>
> (Collins, 2000, p. 254)

African American women's epistemology deconstructs dominant ideologies that justify, support, and rationalize the interests of those in power (Mullings, 1997). Foster (1986, 1989) advocated a leadership that promotes democratic process and calls for political activism that leads to social justice. As a result, this practice of leading is critical, transformative, educative, and ethical. These elements deserve further analysis to identify the important work involved in a critical perspective of leadership.

REFLECTION AND REEVALUATION: CRITICAL LEADERSHIP AND HBCUS

During the period when higher education opportunity was almost nonexistent for African Americans, HBCUs played a significant role in the lives of African Americans who wanted to pursue higher learning. They continue to be a driving force for social change and racial uplift (Jean-Marie, 2004, 2006). HBCUs have a threefold mission: first, they provided education rich in Black history and tradition to newly freed slaves; second, they delivered educational experiences that were consistent with the experiences and values of many Black families; and third, they provided a service to the Black community and the country by aiding in the development of leadership, racial pride, and return service to the community (Sims, 1994). Located throughout the United States, HBCUs symbolize models that educate underserved and underrepresented individuals who traditionally do not have access to institutions of higher learning (Verharen, 1996). Making higher education accessible for every capable individual without compromising quality is an important imperative for HBCUs that continues today (Jean-Marie, 2006).

Verharen (1996) argues that a college education is the right of every competent citizen, and HBCUs pave the way through outreach programs and educational resources to make higher learning obtainable to African Americans. In an era of increased standardization, which aims to limit who should have access to higher learning, other factors beyond traditional criteria (i.e., SAT scores, grade point average, class ranking, financial resources, etc.) ought to be considered to make "quality education available for those who would not otherwise have the opportunity" (Verharen, 1996, p. 53). Among institutions of higher learning, HBCUs seek to bridge the gap between the haves and the have-nots by engaging in critical leadership practices (Jean-Marie, 2006). HBCUs' critical leadership should center on the diverse needs of their student population (financial and educational resources, and other institutional challenges that impact students' success). Many university leaders draw from past experiences to change practices that deter the success of students and progressivism in the Black community.

Beyond universal education for all, HBCUs continue to serve as educational citadels and cultural repositories for the African American community, as well as centers for social and political development of students, faculty and communities, and regions and states in which they are located (Sims, 1994). Although HBCUs were established to serve the educational needs of African Americans, today they serve students from a wide range of cultural and socioeconomic backgrounds. HBCUs not only have racially diverse student populations but many also have a racially diverse faculty and administration. With respect to their enrollment and staff, HBCUs are presently more racially desegregated than historically White institutions (HWIs) (Roebuck and Murty, 1993).

Given the limited research on the significance of HBCUs in higher education, Verharen (1996, p. 53) proposes a renewed charge for HBCUs in their second century. He states:

> HBCUs are strong enough to accept their higher mission: to design models that make radical departures from mainstream education. The radical nature of the problems faced by members of communities that make the existence of HBCUs possible justifies these new models.

In summary, the literature presented here highlights the interconnectedness of socialization, Black feminist theories, and critical leadership as applied to the advancement of women leaders of color. In light of polarity facing women in the workplace, gaining access to and permission to share stories of women leaders of color at HBCUs requires a "great deal of trust . . . because the process of sharing is a political action" (Lawson-Sanders et al., 2006, p. 34). Based on the literature, this approach to generating further understanding represents a fruitful way for researchers to create opportunities to examine more deeply the lives of women leaders at HBCUs and to move beyond valorizing the niceties of feminine values to a deeper embrace of feminism that attends to the experiences of African American female leaders (Benham and Cooper, 1998; Collins, 1990).

RESEARCH DESIGN

Consistent with other research on the use of narrative (e.g., Benham and Cooper, 1998; Casey, 1993, 1995–96; Clandenin and Connolly, 1994; Lawson-Sanders et al., 2006), my study was conducted through the use of narrative inquiry for studying the life stories of African American women. Lawrence-Lightfoot's (1994) use of narrative in her research and writing confirms its suitability for this type of study. She contends that the Black culture is rooted in stories by stating that "a strong and persistent African-American tradition links the process of narrative to discovering and attaining identity . . . and serves as a deep source of resonance" (1994, p. 606). This method assumes that people's lives are stories, and the researcher seeks to collect data to describe those lives (Lawson-Sanders et al., 2006) and provides the means to record and interpret (Riessman, 1993) the voices of women (Benjamin, 1997; Fonow and Cook, 1991; Gilligan, 1982; Gluck and Patai, 1992; Reinharz, 1992). The participants' backgrounds, education, experiences, church, and family that informed their identities were highlighted.

Cooper (1995) and Benham and Cooper (1998) contend that stories speak of the power of narrative in human lives. Cooper (1995, p. 121) maintained that "stories can be retold, reframed, reinterpreted and because they are fluid, open for retelling and ultimately reliving, they are the repositories of hope." According to Benham and Cooper (1998, p. 7), "Narrative methods might very well be more responsive to the researcher's and practitioner's intent to bring to the surface those experiences that go beyond superficial masks and stereotypes." In this study, the stories of the women's interpretations of their leadership practices revealed how they were experts and authors of their own lives. The 12 African American female leaders at HBCUs selected for this study provide a snapshot view of stories of justice, leadership, and racial uplift. As leaders for learning in HBCUs, these women leaders provide a particular scope of the challenges, struggles, and successes they experience.

PARTICIPANTS AND CONTEXT

The data in this study is on 12 African American women leaders at HBCUs in one southeastern state. Pseudonyms identified participants to minimize disclosure of information about individual lives. These leaders served in the capacity of college president, deans of school of education, attorney, vice chancellor, and faculty (a former dean and department chair) at 11 historically Black colleges in the southeastern United States.

Because every text has a context (Casey, 1993), the historical backgrounds of the women's narratives are essential to understanding their self-definitions. These African American women participants grew up during a period in history in which the social and political climates were in upheaval. For some, the "separate but equal" still applied, while others were experiencing the unsettling changes in the early years of desegregation. The expectations of families and members of the African American community for Black young adults to attend college and further their education weighed heavily on these women's shoulders. One participant summed her college experience: "I felt like I was carrying the weight of my race on my back. I constantly felt that I had to do well because if I didn't, I would be letting my people down. It is a big burden to carry."

SEMI-STRUCTURED INTERVIEW QUESTIONS

In the interviews, participants were asked an opening question: "Tell me your life story by reflecting on your personal life in relationship to your professional experiences." The ensuing story, or "main narrative," was not interrupted by further questions but was encouraged by means of nonverbal and paralinguistic expressions of interest and attention, such as "mhm". In the second part of the interview, the "period of questioning," more elaborate narrations on topics and biographical events were discussed. The interviews were one to two hours each; on occasion, another appointment was scheduled to continue the interviews. This was due to time constraints relating to the participants' schedule and professional commitment.

ANALYSIS OF FINDINGS

The key findings that emerged from the data analysis focused on three projects construed from the participants' narratives: a career project rooted in individual achievement, visionary project rooted in economic success, and a social justice project rooted in community. While these areas were dominant in the analysis, several sub-themes materialized within these three key areas.

CAREER PROJECT ROOTED IN INDIVIDUAL ACHIEVEMENT

The group of women who articulated a career project rooted in individual achievement included Presidents Gibson and McGee, Dean Miller, and Attorney Wilson. Only two of these women taught in public schools; Presidents Gibson's and McGee's professional careers started in higher education. Three of the women's educational experiences were from both historically Black and White institutions; Dean Miller obtained all of her degrees from HWIs. To further understand this group of four women whose leadership denotes an individualist style of leading, a discussion on their life experiences segues into their analysis of leadership.

Leading as Individualist: Self-Made Success Stories
It is easy to homogenize the Black woman experience because they share similar interlocking systems of racism and sexism. While the majority of the women openly shared their life struggles, this particular group of four women in the study provided brief narrations about their educational and professional hardships. As I listened to all the voices in the text, I attended to the silences as well. President Gibson, in telling her life story, failed to mention anything about her family:

> I grew up in Benson, NC where I attended public schools. I attended schools during the period of segregation. I grew up picking cotton, barning tobacco, picking up sweet potatoes. And perhaps this hard farm work convinced me to work harder so that I could have a different type of life.
>
> After completing high school, I went on to [HBCU] where I earned a bachelor's degree in (Majors), with an emphasis in teaching. From there I went to [historically White institution] and I earned a master's of science degree in [area of study]. I worked for one year in a rehabilitation hospital as a speech therapist and from there, I went on to work at [HWI] as an instructor in the department of English and Foreign Languages. After spending two years at [HBCU], I then went to [HWI] where I earned a doctorate of philosophy degree in [area of study].

Gibson's story was heavily encoded and she emphasized being oppressed by conditions, not by people. She does not address why she had to perform hard labor nor does she provide an elaborate discussion on her schooling experience during segregation. For Gibson, the past bears ironic significance to where she is today as if to suggest that present-day accomplishments erase past educational and professional hardships.

Like Gibson, Attorney Wilson's narrative presented an individualist orientation. Her childhood experiences disclose why she had to be "self motivated" and "take some ownership" to have arrived this far in her life. For example, she talked about the existential questions she started asking at an early age with no adult present to provide answers to her questions.

> I have always been very much interested in learning. Learning for learning sake as oppose to you go out there, get a degree, get a job. That's fine. But I have always used books as a measure to broaden my concept of why I am here on *this planet at this time.*

Although she questioned why she was here "on this planet at this time," she remained silent about the events that were occurring during her childhood years. She spoke in broad terms about her adolescent years. She explained,

> At a very young age, I had *questions* about why was I here. What was I suppose to do? How do I relate to or handle the kinds of principles that were being instilled that I didn't agree with? What came to mind immediately was my Catholic religion. I resisted the concept that only Catholics are going to go to heaven. Even at a very young age, I knew that didn't make good sense. There's something not right with this. The other thing is that you are encouraged not to question to any great extent.
>   I was most definitely *questioning* in my mind if not openly. So that was a challenge for me. I didn't feel that there wasn't anybody in my family that I could discuss this with. The whole idea about an education is that, it enables you to live a better life. And just information for information sake was not encouraged either by the church or my home. *So that was my challenge.* And I wanted to know about ancient history and how civilization started and how certain practices came into being and that kind of thing.

Attorney Wilson's quest for understanding the world that was revolving around her was not nourished by the two communities she came out of: her family and the Catholic Church. She believed that both institutions, on the contrary, discouraged her from finding truths about knowledge for knowledge's sake, the meaning of God, and what religion really is or what was happening in society. Her family and church seemed to go with the status quo and not challenge any belief systems she called into question. Her resentment toward her two communities' willingness to accept prevailing notions infuriated her to find answers on her own.

What helped, states Wilson, "was being able to read." There lies the answers to the truths she was to learn. The quest became to live life as an individualist because she interpreted her life as one disconnected from church and family ties.

> I read everything in my house and in my great grandmother's house next door. Once I was old enough to go to the library down town, I just went on and on. Eventually I got answers to my questions from books that I read. And when I got to college of course, people that I met, I talked to them.

As she articulated, the challenge for her was "the feeling that I did not have a support group that I could talk to or understood what I was seeing and believing about the world." She felt unsupported or nurtured to help her understand or solve the questions she grappled with as an adolescent and much into her adult life.

> I was very much aware. I think that there are people who know many things and don't know how they know or why they know. I knew that the world around me was changing and I knew that my family was preparing me for a life that was not going to exist at the time and probably didn't exist then.
>   I knew the future was going to be different. Somehow I knew. They [her family] didn't know what the future was going to bring. I knew that there would be a difference. So it sort of takes you outside of your own family when that kind of thing happens.

Her experiences point to her self-discovery about life and its meanings. Wilson summarized her childhood experiences:

> Having a world vision that was not shared by the people around me and I couldn't discuss; it made me an unhappy child. And nobody could say why. You know, "I wasn't being mistreated or anything. I was just different."

As a child who believed she was not nurtured and supported, she survived by relying on her individual efforts to seek and know the world around her. Therefore, she viewed the world through an individualist perspective.

Like Wilson and Gibson, the other women in this group placed emphasis on their upward mobility to highlight personal accomplishments portraying themselves as the "self made" individual.

Authoritarian and top-down styles also characterize the individualist project of leading. Here the leader is the formal authority of power; fosters compliance, not commitment; is unwilling to share leadership; and pays little attention to the organizational history of the institution. An example of this is McGee's views on her leadership style:

> I see myself as a change agent. If there is a situation where they wanted the status quo to be maintained, I am not the least bit interested in it. I like to go in and find a situation that is really in bad shape and fix it. I am ready to go after that.

In making the transition to an HBCU, she failed to consider how Black leadership may operate at this Black women's college. Coupled with paying little attention to the potential Black leadership model that this college operated under, she expressed no interest in learning the cultural traditions of how the university functioned. She envisioned making substantial changes to policies and procedures without considering the resistance she would encounter for failing to include others' voices (i.e., board of directors, others in administration, alumni, community, etc.).

> I never wanted and never said that I would be in a historically black institution just because it was just that. It lacked diversity. It lacked creativity in a number of ways.

It was much too structured, too rigid and it was a monopoly. And I just didn't like that.

But this job came along; I saw it as an opportunity to make the best of the opportunities by succeeding in predominately White environments. If we are to make the kind of impact that we want to make for our children and others then I can bring the experiences that I have to bear in a place like [black women's college] and make it not just a historically black institution, but an institution that is on equal footing with any institution such as *Smith* and other historically White institutions.

President McGee's style of leadership orients itself toward conventional approaches to leading. Her approach of leading came across as inaccessible to the people she worked alongside, such as board members, administrators, faculty, staff, students, and alumni. She failed to take into consideration the existing organizational culture.

Attorney Wilson talked about her leadership approach:

I think for me, leadership means that you are able to identify an issue or problem, to have concepts about solutions, to motivate other people to view the issues the way you do, to look at the solutions the way you do and work toward a specific objective . . . And essentially my style is, if there is one is to try to identify people who are *self motivated* . . .

And over the years, I have been aware of my style of dealing with employees. In private practice, you are hiring people at various positions. And I was very much aware of my *being successful* when I had somebody who was *self motivated*. I was always looking for the person who could *explain the process*, who felt who could *take some ownership* in the position. And so they owned the outcome. And those are the people that I work with best. And I think that I am still that way. Though I accept that there are people who need hand holding and if you probably hold their hand long enough, they might become independent. But I'm not certain of that because I haven't held hands long enough. Either you can do it or you can't.

Similarly, President Gibson describes her leadership in the context of her exceptional abilities as a leader:

If I had to describe my style, I would say that I am extremely organized. I think that I have a strong work ethic to the point that I would work night and day to accomplish a goal. I'm extremely focused. High energy, I probably do the work of about 10 people. I think that I'm very fair with people. Sometimes I think that I may be too loyal to the point that it has hurt me sometimes. I love seeing something accomplished when other people say, it can't be done. And I believe that you can turn the ordinary into the extraordinary with enough hard work and effort.

Drawing from her experiences in a large urban school district to her former role as a dean, Miller approached her leadership from this perspective:

Often times, I bring people in and want them to initiate things. I want them to be leaders but they wait for me to bring ideas to the fore. I challenge them to bring ideas in the concept stage in our discussions. I try to bring people who have a common vision and we know where we are going. I don't want to micro-manage people.

A career model rooted in individual achievement emphasizes the individual leader who practices a top-down and authoritarian leadership. The university moves forward through the actions taken by the leader. An individual leads through a singular perspective and voice that influences policy decision-making and authenticates the leader as the expert to drive the mission of the school.

Inner Circles of the "White Male Model" of Success

Another important distinction of the career project rooted in individual achievement is associated with whom the African American leaders identify as their role models. While one administrator talked extensively about her self-made successes and ascendancy, the other Black women in this group attributed their professional accomplishments to the successful adoption of the "White male model of success." President McGee stated,

> The modeling that I have used has been that of White men. White men seem to have this sense of privilege. They just believe the world is theirs. They believe they can do anything they want. And they don't accept no for an answer. So rather than viewing myself as a Black woman in a predominantly white world, I view myself as a person who is entitled to everything that is out there. I have been very, very stubborn, steadfast and bold in terms of demanding those things that I think I have been entitled to have.
>
> I even got my kicks by succeeding in a predominantly white environment because that was sort of against the norm. So rather than limit myself to this closed structure of being with Afro-Americans and trying to be a big fish in a little pond, I was willing to broaden my parameters and be a small fish in a large pond with the intent of becoming a large fish in a large pond. That's the way I viewed things.

Being socialized into the power dynamic of White men, McGee saw it advantageous to run within a White male circle to achieve recognition, status, and power. As she explained,

> Along the way, it became clear to me that if I was going to reach my goals of being president, I would have to have the doctoral degree. My mentors have generally been White men. And that goes back again to my saying my role models have been White men. And they have been people who have for one reason or another liked my goals and tried to find ways to help me achieve my goals. And they have been very supportive. I have not had the same level of support from White women. And that may be because of the competition factor. It may be a composite of things.

Similarly, Dr. Miller, a former dean of education, was also mentored by White men in pursuit of her professional goals:

> A very interesting thing about my career is that I ended up in the middle of a white male network as an African-American female. It started from my dissertation stage where my advisor of course was White—that's not unusual.
>
> Later, the superintendent in Fairfax told me, "Look we will work out something so that you can take this job. I hate to see you miss this opportunity." He worked it out so I could pay back my time by consulting with the school district rather than

working for them. I then got called to Savannah, Georgia by a *White male superintendent* who said I want to get to know you. He hired me in my first senior level job as assistant superintendent for a psychological counseling and services in special education.

The participants seem to suggest that they seek career opportunities rather than communal ties. They temporarily take on positions for very short periods until they are ready to move to the next level. President McGee and Dean Miller talk extensively about how within short periods they changed jobs. Miller spoke about her concern of being pigeonholed into a position:

> I had become somewhat career-oriented, still not very ambitious but career oriented and knowing where I wanted to be and what I wanted to do. I made a deliberate decision that I was going to keep my eye out for my ideal position and low and behold it became available and I applied.

She continued, "There has been a pattern in my career of every three years I got a promotion." President McGee also strategically planned her move to become a college president:

> A lot of times we think about doing strategic planning for our business and for our schools, but we don't do strategic planning for our lives. We don't think about our mission. We don't think about what our goals and objectives are. We don't think of the resources that we need to accomplish them. I did. So I said, "If my goal is to become a president then I want to do it within three to five years. Therefore, I needed to get busy".

In summary, the women leaders in this group spent a great deal of time talking about their upward social mobility and self-made successes by modeling White men. They articulated the traditional, authoritarian leadership as their model of practice. Evident from this group of women's narrative is that they have adopted an individualist style of leadership. Their orientation is attributed to the lack of institutional or communal ties to people in their professional environment. For example, as individualist, one places little interest in knowing the existing culture of the institution. There is little effort in getting to know the people to understand what the problems and issues the institution faces; instead, the individualist assumes to already know the solutions before accurately assessing what the problems are. As President McGee explains,

> I like situations *where there are problems* and *I know the answers* to the problems. I have the solutions already. And in some cases I have taken jobs without having a clue of how I was going to accomplish the responsibilities.

VISIONARY PROJECT ROOTED IN ECONOMIC SUCCESS

The women who articulated a visionary project rooted in economic success included President Murphy and Dr. Royster. President Murphy's work experiences are within public schools and higher education; she pursued her educational degrees at both historically Black and White institutions. Dr. Royster's work experiences are from the corporate world and higher education; her educational degrees have all been at HWIs.

A visionary project rooted in economic success is based on a corporate mindset operating in higher education. The women leaders identified in this group have prior experiences in the business sector and transmit the characteristics of the business culture into the academic culture. Similar to the corporate world that operates from the concepts of products, visionary purposes, efficiency, and delivery of goods and services, this model viewed students as "consumers" whose education is the product being sold.

Drawing from her experience in the business sector, Dr. Royster, an academic vice chancellor, approached her leadership practices through an economic mindset. Articulating her vision for her university, she asserted,

> I want young people and myself to understand the business aspects of what we do because this is a business. *The business is education.* You have to understand the competitive forces out there in the environment. If I want to grow my enrollment, I've got to be competitive. To be competitive, I've got to have first class products. I have to deliver services in a first class way.

To deliver the goods, she placed value in collaborating with her colleagues, staff, and students. They were driven by success and as Royster claimed, "You have to be about excellence day in and out as partners in this whole enterprise." Speaking most passionately about her hopes for her students, Royster spoke in a forthright manner about the difficulties many of them experience. She believed that the "diamonds in the rough" can succeed in their pursuit of higher education through mentorship and academic preparation.

Students as Consumers: Economic Conservation of the Black Community

Expressing similar concerns to the social justice project, President Murphy and Associate Vice Chancellor Royster interrogate the implications of African Americans' lack of educational preparation in schools. Both participants' leadership approaches are from an economic perspective that fulfills what I call "the Black economic project," intended to equip students with the requisite skills to function in and out of their communities.

Preparing students to achieve their personal and professional goals was a central concern to the women in the economic project. They had a particular concern regarding the overall development of their students that dates back to integration. In the early years of her teaching career in the South, President Murphy's path to fight inequities and injustices was carved out soon after she graduated from college and started her first year of teaching:

> It was the first year [city] had mass busing and I taught in an elementary school. It was set on a hill in the middle of an all white neighborhood and they [school officials] bussed kids from the poorest section of [city] to one of the more affluent communities in [city]. It was a more unlikely match you could have made. Those [African-American] kids endured a horrendous and horrific year. I was the youngest teacher on staff by 23 years and I was the only one of four African Americans.

I watched what was happening to those kids who didn't have anyone to stand up for them. I then realized what happens to people who look like me but who didn't grow up charmed and have an advocate to address their injustices . . . It was at that point I realized there was a whole population of people that didn't have an advocate.

Witnessing the struggles of disadvantaged children in the African American community, Murphy felt compelled to be the lonesome voice and risked her job to challenge the daily inequities she witnessed. Despite her failed attempts to change the conditions of these students while a teacher at the school, she became more determined to be a voice for underrepresented children and has done so throughout her years in public and higher education.

One initiative of Dr. Murphy, a newly appointed first female president of one private HBCU, was a policy decision that would impact many students' educational experience. Although it was a major undertaking and had resistance from her governing board and staff, President Murphy developed a plan to attract African American students who ordinarily would not be admitted to her private university or any other higher education institution (i.e., low SAT scores and grade point average; limited to no financial resources). Her approach was as follows:

Next year, we're going to start what we're calling a pilot program and admit about 150 of our freshmen class. They will be kids who may not have good SAT scores or have all the credentials, but somebody in that community whether it is in the school or the church says, "This is a kid to take a chance on. Trust me. Take a chance on this kid." We are going to admit this kid and introduce them to a curriculum that is outside the box . . . We are going to prove that by staying close to the mission—that is, if you provide access to education for kids, SAT scores mean absolutely nothing. It is potential that makes a difference. That is the real predictor of how well kids are going to succeed.

Garnering the support of the local community, schools, and churches, she identified and admitted 150 students who ordinarily would not have attended college. Understanding that these students would confront numerous challenges to complete their academic studies, she developed a five-year program that included year-round academic support, remedial classes, financial resources, and mentoring for the students she admitted. Murphy asserted, "We're going to take a kid who nobody else thinks they ought to take a chance on except some momma, some daddy, some preacher, some teacher or some faculty member. We're going to educate them."

For both of these leaders, the plight of the African American community was a social responsibility they did/could not abandon; it was imperative that they engaged in oppositional discourses to challenge the status quo of that time period. In sum, the two leaders oriented in an economic project interpreted their leadership practices as fulfilling an important purpose, not only for their respective schools but also for the development of the Black community. Their firm commitment to all students who attended their universities is packaged through visionary leadership, which is intended to equip students with the necessary knowledge and skills to navigate in and out of their communities. The women leaders drove the vision and set the directions that prepare students to go out and compete with others in the "market," which is society.

SOCIAL JUSTICE MODEL ROOTED IN COMMUNITY

The group of women who belonged to a social justice project rooted in community included Deans Frazier and Smith and Drs. Giddens, Johnson, Allen, and Owens. All of them except Dean Smith taught in public schools prior to their experience in higher education. Dean Smith's background is in the field of nursing. Four of the women's educational experiences were both at historically Black and White institutions; Dr. Giddens attended only historically Black institutions to pursue her studies, and Dr. Owens obtained her degrees from HWIs.

The social justice project is consistent with an interpretive tradition founded in both Black churches and historically Black institutions, which focus on social justice through shared work. The leaders in the social justice group interpreted their leadership in the context of a spiritual/religious realm and through shared community commitment. They regarded students as children of the community who are to be nurtured, groomed, and prayed for as leaders of the next generation.

Preparing Underserved Students to Achieve Success

Preparing students to achieve their personal and professional goals was a central concern to the women in this project. Collectively, these leaders viewed their students as the children of the community. They had a particular concern regarding the overall development of students. Dr. Giddens, a former department chair and dean, talked about the struggles of many of her students:

> Since I've come there, it has always been a struggle to develop students because of the students that we serve and the environment—lack of equipment and resources, trying to work with faculty members who don't care [and] students [who] don't have initiative and drive. That has been a struggle to truly do that. It has been a struggle for me to even remain at (HBCU), knowing the potential that I have, but I stay there for the students. I truly *love* the students here.

According to Dean Frazier, school of education,

> My students at this university are students I know that most other universities would not touch. Not because they don't have the ability, because they do but you have to reach inside and pull that ability out. But because [college personnel] have such high standards, they feel like "Well we just don't have to bother with that student." That's the kind of student that we have here. We have all kinds of students here. We also have students that have had all kinds of experiences, and I believe that there is an opportunity to give credit for a lot of experiences.

The participants also expressed concern about the lack of academic preparation students receive prior to matriculating to higher education. Many HBCUs are aware of their students' lack of academic preparedness and seek to provide needed resources to help them. As Dean Frazier asserted, "Bring them into that village called [HBCU] and we will work with them!"

A social justice model rooted in community accentuates the interpretive tradition of HBCUs. Community building coincides with the mission of preparing communities of Black scholars academically, culturally, and socially. Similar to many Southern Black women leaders, this group of women viewed their vocations as service to God and to community.

Not only were they actively involved in policy decision-making that affected the academic curriculum, they were also committed to providing a moral and social curriculum to ensure their students' success. In Bakhtin's (1981) phrase, they exercise "response-ability" as "village elders." In essence, they become organic intellectuals of and for the community.

Important to this group were their religious/spiritual relationships that operated on two dimensions. The first was the personal relationship with God, which the Black women leaders considered the source of their strength to help them navigate with the world. Dr. Johnson, an academic VP, spoke of how she came into her present position on campus:

> From the very first day, I just knew this was the place for me and I know that God sent me there. There is no question because I had been praying. I said, "God, wherever you want me to go, that's where I will go."

Frazier spoke extensively about her spiritual existence as an indispensable aspect of her personal and professional life:

> My spiritual connection gives me some stability. And everyday of my life I think about what that is. And recently I thought for example with my parents, there is a spiritual piece there to them that moved me to the point where I am now even though they're deceased. There is still that spiritual connection with them that we don't quite understand. But I think that influence is there to guide us along the way.

The second dimension involved their social relations developed through their church. The women administrators identified here had strong ties to their churches, either through their own ministry or through active involvement in their church community. They brought these values into their work environment.

Finally, their social justice project spoke to their commitment to uplift the African American community. As part of their purpose, the women were committed to developing African Americans to be academically competent in their chosen profession and to be culturally knowledgeable and socially attentive to their own community as well as to the larger society. The Black women leaders' narratives in this group depicted how their professional work intersects with their spirituality and commitment to "uplift the race." Whether in their formal professional roles as educators in public education or in their recent roles as leaders in HBCUs, the participants articulated a leadership that is tied to social change, institutional reform, and structures and processes of power and influence—what the Civil Rights movement was about. Their involvement and ongoing interactions with students, staff, and constituents at their institutions and community characterize a social and political activism that is reminiscent of leadership practices of their predecessors of the Civil Rights Movement.

CONCLUDING THOUGHTS: INTERPRETIVE TRADITIONS AND COMMUNITIES

Although the women interviewed were situated in HBCUs, they do not all embrace the values ascribed in the mission of historically Black institutions. Like Black churches, HBCUs are situated in the heart of African American neighborhoods throughout the United States. They symbolize a sense of hope and aspiration and place of refuge for Blacks. They also have been pivotal in revolutionary movements such as the social and economic development of Blacks. Therefore, the collective project of social justice is an element of the interpretive tradition in historically Black institutions. Interpretive traditions refer to vocabulary, metaphors, accepted wisdom, and norms used by persons with similar experiences. These experiences form a common location in which members of the group become interpretive communities whose ways of being reflect their connectedness (Jean-Marie, 2003; Rutherford, 2001; Still, 1994).

Through associations, storytelling, and other forms of communiqués groups have the opportunity to add other interpretive traditions to their repertoire of language when their conversations are with persons from other traditions. The career project rooted in individual achievement and the visionary project rooted in economic success bear evidence of other interpretive traditions the Black women leaders in the two groups have adopted in their social repertoire. African American women leaders whose experiences draw from other interpretive traditions and have adopted those traditions as their repertoire of social languages and practices confront opposition in institutions with an interpretive tradition that values shared commitment and collective work (Jean-Marie, 2003). Much of the challenges have to do with their attempts to transmit HWIs' interpretive tradition to historically Black institutions.

Of the three projects, the social justice project rooted in community is most closely aligned with the African American interpretive tradition of community building. Their practices of leadership serve to transform existing conditions and empower those they serve, including students, faculty, staff, and community. At the heart of what they do is facilitating democratic participation.

The female leaders communicate to institutional members through discussions and actions that change can be made by an "us" mentality. Though there are many struggles such as lack of funding and other resources, the work that is being performed has larger implications that go beyond institutional purposes. Their political engagements and conscious efforts to transform the educational conditions of African American students speak to a concern with ideologies and activities in developing specific groups or communities. Finally, the religious/spiritual faith of these women has been their guiding force to lead, to educate themselves and others, and to change repressive conditions. As Collins (1998) avers, justice constitutes an article of faith expressed through deep feelings that move people to action. For these administrators, their concern with justice infused with a deep spirituality move them to build communities of socially responsible students in continuation and preservation of the African American culture.

Looking back on their life's journey, the participants shared the significance of their personal and professional experiences. An important message conveyed by these Black women leaders to members within and outside academia is that they, as leaders, are not to be discounted in working toward social change. For many of them, HBCUs are where they are supposed to be and can have the most impact on students, administrative decisions, and educational reforms. Although many of their struggles go unrecorded, only when these stories are told can individuals within and outside the Black community understand the sacrifices of African American women who fought against injustices and continue to pave the way for generations to follow their lead. Understanding their purpose, they stand firm on the promise to lead HBCUs to a level of excellence despite the challenges of students and institutional practices and policies.

REFERENCES

Bakhtin, M. M. 1981. *The dialogic imagination*, Ed. M. Holquist. Trans. C. Emerson and M. Holquist. Austin, TX: University of Texas Press.

Bass, B. M., and B. J. Avolio. 1994. Shatter the glass ceiling: Women may make better managers. *Human Resource Management* 33: 549–60.

Benham, M., and J. Cooper. 1998. *Let my spirit soar: Narratives of diverse women in school leadership*. Thousand Oaks, CA: Corwin Press.

Benjamin, L. 1997. *Black women in the academy: Promises and perils*. Gainesville, FL: University Press of Florida.

Bennis, W. 1985. *Leaders: The strategies for taking charge*. New York: Harper & Row.

Brunner, C. C. 1998. Women superintendents: Strategies for success. *Journal of Educational Administration* 36(4): 160–82.

Caprioli, M., and M. A. Boyer. 2001. Gender, violence and international crisis. *Journal of Conflict Resolution* 45: 503–18.

Casey, K. 1993. *I answer with my life*. New York: Routledge.

Casey, K. 1995–96. The new narrative research in education. *Review of Research in Education* 21: 211–53.

Clandenin, D. J., and F. M. Connolly. 1994. Personal experience methods. In *Handbook of qualitative research*, ed. N. K. Denzin and Y. S. Lincoln, 413–27. Thousand Oaks, CA: Sage.

Collins, P. H. 1990. *Black feminist thought: Knowledge, conscious, and the politics of empowerment*. New York: Routledge.

———. 1996. The social construction of Black Feminist thought. In *Women, knowledge, and reality: Explorations in Feminist philosophy*, ed. A. Garr and M. Pearsall. New York: Routledge.

———. 1998. *Fighting words: Black women and the search for justice*. Minneapolis, MN: University of Minnesota Press.

———. 2000. *Black feminist thought: Knowledge, consciousness, and the politics of empowerment*. New York: Routledge.

Cooper, J. E. 1995. The role of narrative and dialogue in constructivist leadership. In *The constructivist leader*, ed. L. Lambert, D. Walker, D. P. Zimmerman, J. E. Cooper, M. D. Lambert, M. E. Gardner, and P. J. F. Slack, 121–33. New York: Teachers College Press.

Cubillo, L., and M. Brown. 2003. Women into educational leadership and management: International differences? *Journal of Educational Administration* 41(4): 278–91.

Davis, A. 1989. *Women, culture and politics*. New York: Random House.

DeMatteo, L. A. 1994. From hierarchy to unity between men and women managers. Towards an androgynous style of management. *Women in Management Review* 9: 21–8.

Dorn, S. M., C. L. O'Rourke, and R. Papalewis. 1997–98. Women in education: Nine case studies. *National Forum of Educational Administration and Supervision Journal* 14(2): 13–22.

Eagly, A. H., S. J. Karau, and B. T. Johnson. 1992. Gender and leadership style among school principals: A meta-analysis. *Educational Administration Quarterly* 28(3): 76–101.

Evans, G. A. 2001. The world on our backs. *Community College Journal of Research and Practice* 25: 181–92.

Feldman, D. C. 1989. Socialization, resocialization and training: Reframing the research agenda. In *Training and Development in Organizations*, ed. I. L. Goldstein, Irwin Goldstein, and Sybil Goldstein, 376–416. Newbury Park, CA: Corwin Press.

Freeman, S., and R. Varey. 1997. Women communicators in the workplace: Natural born marketers? *Marketing Intelligence and Planning* 15: 318–24.

Furman, G. C., and R. J. Starratt. 2002. Leadership for democratic community in schools. In *The educational leadership challenge: Redefining leadership for the 21st century*, ed. J. Murphy, 105–33. Chicago: National Society for the Study of Education.

Fonow, M. M., and J. A. Cook, eds. 1991. *Beyond methodology: Feminist scholarship as lived research*. Bloomington, IN: Indiana University Press.

Foster, W. 1986. *Paradigms and promises: New approaches to educational administration*. Buffalo, NY: Prometheus Books.

———. 1989. Toward a critical practice of leadership. In *Critical perspectives on educational leadership*, ed. J. Smyth, 39–62. New York: Falmer.

Garibaldi, A., ed. 1984. *Black colleges and universities: Challenges for the future*. New York: Praeger.

Gilligan, C. 1982. *In a different voice: Psychological theory and women's development*. Cambridge, MA: Harvard University Press.

Gluck, S. B., and D. Patai, eds. 1992. *Women's word: The feminist practice of oral history*. New York: Routledge.

Greenleaf, R. 1996. *On becoming a servant leader*. Indianapolis, IN: Robert K. Greenleaf Center for Servant-Leadership.

Grogan, M. 1994, April. *Aspiring to the superintendency in the public school systems: Women's perspectives*. Paper presented at the Annual Meeting of the American Educational Research Association, New Orleans, LA.

Gronn, P. 2003. *The new work of educational leaders: Changing leadership practice in an era of school reform*. London: Paul Chapman.

Hine, D. 1994. *Hinesight: Black women and the re-construction of American history*. Brooklyn, NY: Carlson.

hooks, b. 1981. *Ain't I a woman: Black women and feminism*. Boston, MA: South End Press.

———. 1989. *Talking back: Thinking feminist, thinking Black*. Boston, MA: South End Press.

———. 1990. *Yearning: Race, gender, and cultural politics*. Boston, MA: South End Press.

———. 1994. *Teaching to transgress: Education as the practice of freedom*. New York: Routledge.

Jean-Marie, G. 2003. The courage to lead: Black women administrators speak. *Women in Higher Education* 12(6): 18–20.

———. 2004. Black women administrators in historically Black institutions: Social justice project rooted in community. *Journal of Women in Educational Leadership* 2(1): 37–58.

———. 2005. Standing on the promises: The experiences of Black women administrators in historically Black institutions. *Advancing Women in Leadership Online Journal* 19: 1–14. Available at http://www.advancingwomen.com/awl/fall2005/19_3.html.

———. 2006. Welcoming the unwelcomed: A social justice imperative of African-American female leaders. *Educational Foundations* 20(1–2): 83–102.

Klenke, K. 2003. The "S" factor in leadership: Education, practice and research. *Journal of Education for Business* 79 (1): 56–60.

Kluckhohn, C. C. 1951. The study of culture. In *The policy sciences*, ed. D. Lerner and H. D. Lasswell, 85–97. Stanford, CA: Stanford University Press.

Ladson-Billings, G. 1997. For colored girls who have considered suicide when the academy's not enough: Reflections of an African American woman scholar. In *Learning from our lives: Women, research, and autobiography in education*, ed. A. Neumann and P. L. Peterson, 52–70. New York: Teachers College Press.

Lawrence-Lightfoot, S. 1994. *I've known rivers: Lives of loss and liberation.* Reading, MA: Addison-Wesley.

Lawson-Sanders, R., S. Campbell-Smith, and M. K. P. Benham. 2006. Wholistic visioning for social justice: Black women theorizing practice. In *Leadership for social justice: Making revolutions in education,* ed. C. Marshall and M. Oliva, 31–63. Boston, MA: Pearson Education.

Loder, T. L. 2005. On deferred dreams, callings, and revolving doors of opportunity: African-American women's reflections on becoming principals. *Urban Review* 37(3): 243–65.

Morris, A. D. 1994. *The origins of the civil rights movement: Black communities organizing for change.* New York: Free Press.

Mullings, L. 1997. *On our terms: Race, class and gender in the lives of African American women.* New York: Routledge.

Normore, A. H. 2006. From a pivotal Civil Rights activist to radical equations: Grassroots leadership and lessons for educational leaders. A conversation with Robert Moses. *University Council of Educational Administration Review* 48(1): 19–22.

Oakley, J. G. 2000. Gender-based barriers to senior management positions: Understanding the scarcity of female CEOs. *Journal of Business Ethics* 27: 321–34.

Olsson, S., and R. Walker. 2003. Through a gendered lens? Male and female executives' representations of one another. *Leadership and Organization Development Journal* 24: 387–96.

Pounder, J. S., and M. Coleman. 2002. Women–better leaders than men? In general and educational management it still "all depends". *Leadership and Organization Development Journal* 23: 122–33.

Reinharz, S. 1992. *Feminist methods in social research.* New York: Oxford University Press.

Riessman, C. K. 1993. *Narrative analysis.* Newbury Park, CA: Sage.

Robnett, B. 1997. *How long? how long?: African American women in the struggle for Civil Rights.* New York: Oxford University Press.

Roebuck, J. B., and K. S. Murty. 1993. *Historically Black colleges and universities: Their place in American higher education.* Westport, CT: Praeger.

Rutherford, S. 2001. Organizational cultures, women managers and exclusion. *Women in Management Review* 8: 371–82.

Schein, E. H. 1992. *Organizational culture and leadership.* 2nd ed. San Francisco, CA: Jossey-Bass.

Schiller, N. 2000. A short history of Black feminist scholars. *Journal of Blacks in higher education* 29: 119–24.

Shuttleworth, D. 2003. *School management in transition: Schooling on the edge.* London: Routledge Falmer.

Sims, S. J. 1994. *Diversifying historically Black colleges and universities: A new higher education paradigm.* Westport, CT: Greenwood.

Still, L. 1994. Where to from here? Women in management. The cultural dilemma. *Women in Management Review* 9: 3–10.

Van Engen, M. L., R. Van der Leeden, and T. M. Willemsen. 2001. Gender, context and leadership styles: A field study. *Journal of Occupational and Organizational Psychology* 74: 581–98.

Verharen, C. 1996. Historically Black colleges and universities and universal higher education. In *Ethics, higher education and social responsibility,* ed. J. A. Ladner and S. Gbadegesin, 45–58. Washington, DC: Howard University Press.

Wesson, L. H., and M. L. Grady. 1993, March. *A comparative analysis of women superintendents in rural and urban settings.* Paper presented at the National Conference on Creating the Quality School, Oklahoma City, OK.

Williamson, R. D., and M. B. Hudson. 2001, April. *New rules for the game: How women leaders resist socialization to old norms.* Paper presented at the Annual Meeting of the American Educational Research Association, Seattle, WA.

Young, M. D., and S. McLeod. 2001. Flukes, opportunities, and planned interventions: Factors affecting women's decisions to become school administrators. *Educational Administrators Quarterly* 37: 462–502.

PART 2

TROUBLES

McCarthyism's Effect on Black Colleges in Pennsylvania: A Historical Case Study of Cheyney and Lincoln Universities

PATRICIA C. WILLIAMS

The Black colleges were founded and developed in an environment unlike that of other colleges—one of legal segregation and isolation from the rest of higher education. These colleges served a population that lived under severe legal, educational, economic, political, and social restrictions (Hill, 1985). In addition, since there was no system of public education for Blacks in the South, where 94% of the Black population lived, the establishment of elementary schools, secondary schools, and colleges at the same time created a unique set of challenges that had to be addressed following the Civil War (Drewry and Doermann, 2001). Thus, the system of Black colleges and universities in America can be attributed largely to two events: the North's defeat of the South during the Civil War and the Morrill Land Grant Act of 1890.

Cheyney University was established in 1837 with the help of Richard Humphrey, a Quaker philanthropist who witnessed race riots in Philadelphia in 1829. Humphrey left $10,000 for the design and establishment of a school to educate descendants of Africa for positions as teachers. He left this charge to 13 fellow Quakers who began the school as the Institute for Colored Youth and offered free classical education (www.cheyney.edu). Cheyney University is the oldest of the Black institutions in the United States. Although Lincoln University also claims this honor, the General Assembly of Pennsylvania recently declared, "Whereas, Cheyney University of Pennsylvania holds a singular place in our nation's history as America's oldest historically Black institution of higher education . . . (House Resolution

No. 603, 2006)," thereby affirming Cheyney as the first Black college. Following the 1902 move to the farm of George Cheyney, the institution underwent several name changes until in 1959 eight years after accreditation by the Middle States Association of Colleges and Secondary Schools it became the Cheyney State College. Cheyney joined the new State System of Higher Education and became Cheyney University of Pennsylvania in 1983 (www.cheyney.edu).

Also located in Pennsylvania, Lincoln University was founded by John Miller Dickey and his wife, Sarah Emlen Cress, as the Ashmun Institute in 1854. The institution became Lincoln University in 1866 when it was renamed after President Abraham Lincoln. In 1952 Lincoln University opened its doors to women. The institution joined the state system in 1972 as Lincoln University of the Commonwealth of Pennsylvania. Like Cheney University, Lincoln is accredited by the Middle States Association of Colleges and Schools (www.lincoln.edu).

FACING CHALLENGES

In 1938 Congress fueled its own growing distrust of domestic Communism by creating the House of Representatives' Special Committee on Un-American Activities. This committee initiated the very tactics of secretive investigations, finger-pointing, unsubstantiated accusations, and public inquiries that Senator Joseph R. McCarthy and his followers employed during the late 1940s to mid-1950s to "expose" Communists (Hutcheson, 1997). In fact, the senator was so successful at his task of seeking out and identifying Communists within the ranks of the federal government, the work place, the entertainment industry, as well as in colleges and universities—whether or not they were an actual threat to national security—that most Americans refer to this time period as the McCarthy Era, and to the practice as McCarthyism. Hutcheson (1997) offers a more detailed explanation:

> Advocates of McCarthyism used lack of evidence as a means to attack and as a means to sustain attacks and sanctions among professors. McCarthyism includes the widespread denial of appointment and promotion (or even the retention of position), with lists of the damned. McCarthyism includes the fear of professors and administrators, identifiable for example in their testimony when they identified their colleagues as Communist sympathizers in order to avoid incrimination . . . Finally, it also consists of the variable use of institutional procedures to confirm external accusations.
>
> (p. 619)

McCarthy personally investigated only one professor, Owen Lattimore, the director of the Page School of International Relations at Johns Hopkins University during Communist China's overthrow of Chiang Kai-Shek (Hutcheson, 1997). Lattimore, who had also served as a personal advisor to Chiang Kai-Shek (Freedom of Information Act on Owen Lattimore [foia.fbi.gov/owenlatt1apdf]), was accused by two informants of being a Communist and a Soviet agent. Following a year-long investigation during which 19 informants were interviewed (pp. 150–1), in February 1950 Senator McCarthy presented evidence against Owens and eight others to the Tydings Committee—a subcommittee of the Senate Foreign Relations Committee that Congress established to investigate whether or not the State Department had been infiltrated by Communists (*New American,* May 11, 1987). The committee cleared all nine of the accused and dismissed McCarthy's charges as a "fraud" and a "hoax" (*New American,* May 11, 1987, p. 5). In spite of this setback, Senator McCarthy remained instrumental in promoting a "mass hysteria" that swept the country (Hutcheson, 1997, p. 465). According to Lionel Lewis (1988), McCarthyism resulted in attacks upon the principles of academic freedom and professors at 58 American colleges and universities between the years 1947 and 1956. In *The Academic Mind: Social Scientists in a Time of Crisis,* Lazarsfeld and Thielens (1958) found an estimated 990 incidents at 165 institutions of higher education where professors had been asked to sign loyalty oaths, had lost promotions, and/or were fired because of alleged Communist affiliations. However, even prior to McCarthy's reign, other politicians had already accused colleges and universities of "harboring Communists or former Communists" (Hutcheson, 1997, p. 612).

Nonetheless, the American Association of University Professors' (AAUP) "1940 Statement of Principles of Academic Freedom and Tenure" assures the professoriate of their freedom as private citizens:

> When they speak or write as citizens, they should be free from institutional censorship or discipline, but their special position in the community imposes special obligations. As scholars and educational officers, they should remember that the public may judge their profession and their institution by their utterances. Hence, they should at all times be accurate, should exercise appropriate restraint, should show respect for the opinions of others, and should make every effort to indicate that they are not speaking for the institution.
>
> (Part C)

In spite of the statement, many college and university professors, as well as administrators, were forced to sign loyalty oaths and were intimidated by the boards of regents, trustees, or the presidents of the school into testifying against colleagues or face the possibility of losing their jobs or tenure. These occurrences have been well documented by scholars writing about McCarthyism at historically White institutions (Kille, 2004; Hutcheson, 1997; Schrecker, 1986; Diamond, 1992; McCormick, 1989; Lewis, 1988), but what about Black colleges and universities? A survey by R. Grann Lloyd (1952) of 104 Black colleges revealed that ten of the 78 respondents required their employees to sign loyalty oaths. All ten were public institutions. One school, Prairie View State College in Texas, also required its students to sign such oaths. Therefore, this study indicates that the powerful forces of McCarthyism were also present at Black institutions and that they affected not only faculty and administrators but students as well.

Thus, as both Schrecker (1986) and Hutcheson (1997) agree, while there is an "assumption of power" of the professoriate (Schrecker, 1986), academic freedom is ambiguous. This ambiguity is especially evident in the 1953 statement by the AAUP, "The Present Danger," which affirmed that institutions should investigate those accused of being Communists, as well as those who invoked the Fifth Amendment (Hutcheson, 1997). Therefore, professors, who once thought they were protected under academic freedom, now found their careers threatened because of their political affiliations and private lives. However, because McCarthyism delved into the private lives of the professoriate, it exemplified not only the ambiguities of academic freedom for faculty, but also the vulnerability of the administrators of colleges and universities who were responsible for upholding or disavowing these principles (Hutcheson, 1997) and who were often faced with making difficult decisions based upon the political climate and the needs of their institutions.

According to Hutcheson (1997), the susceptibility of many schools to McCarthyism could be attributed more to financial and fund-raising concerns and the pressures of local constituencies than to the national political climate. This was certainly the case at Fisk University during Charles S. Johnson's tenure from 1947 to 1956, the heart of the McCarthy period. Johnson had to choose between upholding professor Lee Lorch's academic freedoms or making a decision that ensured the continued financial stability of Fisk University, which depended largely upon outside philanthropists. The pressure on Johnson was compounded by the fact that he was the university's first Black president. In her essay, "Scylla and Charybdis: Navigating the Waters of Academic Freedom at Fisk University during Charles S. Johnson's Administration," Marybeth Gasman (1999) defends Johnson's decision to dismiss the Northern White professor Lee Lorch for his previous affiliation with the Communist party and his civil rights activism, which offended White Southerners in Tennessee. As a historian of philanthropy at Black colleges, Gasman sees Johnson's decision as a justifiable one if Fisk University were to continue to receive government grants and private funding from its Southern benefactors. However, she raises several interesting questions for historians of higher education to address. First, how did McCarthyism affect the Black colleges and universities? And second, were there other incidents of infringement upon academic freedom, which affected professors or administrators at other Black colleges and universities? Since, as Hutcheson (1997) states, McCarthyism represented "attacks on those who were different—whether their differences were religious, ethnic, or sexual" (p. 615), one may conclude that such differences included race as well. As a result, Fisk and other Black colleges were particularly vulnerable to the pressures of the McCarthy Era. This defenselessness to the incursions of McCarthyism may have been exacerbated by the lack of funds and dependence upon Northern industrial philanthropists and White Southern donors with their own agendas (Gasman, 1999). Through their control of funding, corporate donors often sought to have power over freedom of speech as well. In fact, even in the early 1940s, prior to McCarthyism, at a time when there was minimal federal interference on college and university campuses, Black institutions, especially those in the South, were subject to reprisals, such as the withholding of funds, from their White benefactors, if they did not conform to expected behaviors (Williamson, 2003). In addition, in the aforementioned study, Lloyd also cites an article in the *Norfolk, Virginia Journal and Guide* (July 16, 1949, p. 1), which suggests that at least one president of a Black institution was fired for his "disapproval of loyalty oaths" (as cited by Lloyd, p. 8, footnote 1). Again, this raises the question: Were administrators at Black institutions more vulnerable to the threats and power of McCarthyism because of their reliance on external funding? Also, with the understanding that faculty and administrators at Black schools were affected by McCarthyism, what then was the climate like for students at Black colleges and universities?

Consequently, with the above questions in mind, and in an effort to expand scholarly discourse that answers Hutcheson's (1997, p. 624) question, "What about stories of those on the margin?" this chapter discusses the impact of McCarthyism on two Black universities in Pennsylvania: Cheyney and Lincoln. It does so by examining several key issues at the two universities during the McCarthy Era: student life and enrollment, funding and endowment, and campus climate.

STUDENT LIFE AND ENROLLMENT AT LINCOLN AND CHEYNEY

From the late 1940s through the mid-1950s the students at these two institutions held varied interests outside of the classroom environment. Thus, there existed student organizations or societies for athletics, governance, literature, music, oratory, philosophy, politics, and religion. Students also expressed a "growing interest in Greek letter organizations" (*Cheyney Record,* March 1951, p. 1). In addition, as a result of the GI Bill of Rights of 1944, Black men returning from war enrolled at Lincoln and Cheyney. Thus, a larger veteran presence emerged on both campuses, as indicated by numerous articles in their respective student newspapers, *Cheyney Record* and *Lincolnian.* Articles such as "Vet Villagers Protest Removal of Stoves" (*Lincolnian,* March 29, 1947, p. 1) and "Lincoln Vets Sponsor Second Spring Dance" (*Lincolnian,* January 22, 1947, p. 1) clearly indicate that veterans were not merely present in terms of numbers, but were active leaders on campus as well. How ironic that these men, who had put their lives on the line during the war, would later be accused of being Communists when they fought for their own freedoms as Black Americans, here at home (Rosenthal, 1975).

However, the colleges differed from White institutions, as well as each other, in the makeup of their student bodies. While Cheyney remained an all-Black institution, with most of its students coming from Pennsylvania and especially the Philadelphia area, Lincoln attracted an interracial and international enrollment (most of the international students were from African countries). As suggested by its website (www. lincoln.edu), Lincoln was where Blacks who sought to become doctors or lawyers, as opposed to teachers, came for an affordable quality education. Although Cheyney's enrollment was consistently lower than that of Lincoln, the college grew during the age of McCarthyism, while Lincoln's enrollment declined during this ten-year period (see Tables 5.1 and 5.2). Because of its dependence upon the enrollment of out-of-state and international students, it is possible that this decrease in numbers at Lincoln was a result of the war, as well as the ensuing threat of Communism.

Private Black schools generated the majority of their revenue from student tuition and fees. However, these institutions also relied heavily on philanthropy and other sources of external funding (Rosenthal, 1975; Williamson, 2003), such as state grants and appropriations. Consequently, these institutions were still subject to gubernatorial expectations for professorial behavior, as well as state approval for the school's curriculum. Failure to comply with state regulations may have meant loss of licensing or certification, which could have resulted in the school's doom (Rosenthal, 1975). Therefore, as was exemplified in the Charles S. Johnson and Lee Lorch case at Fisk University, leaders of Black schools, whether private or state-affiliated, faced great external pressures to comply with the political agendas of their benefactors

Table 5.1   Cheyney enrollments, 1946–1955

Years	Pennsylvania	Out of state	Part-time	Total
1946–47*	–	–	–	280
1947–48**	–	–	–	–
1948–49*	–	–	–	320
1950–51*	–	–	–	370
1951–52	330	46	1	377
1953–54	324	52	23	399
1954–55	365	58	15	438

*Estimated numbers; breakdown not available in bulletin/catalogs for this year
**Bulletin/catalog not found in archives
Source: Cheyney University Archives.

Table 5.2   Lincoln University enrollments, 1946–1955

Years	Pennsylvania	Out of state	International	Total
1946–47	176	377	31	584
1947–48	211	355	30	596
1948–49	184	324	23	531
1949–50	178	272	33	483
1950–51	172	213	39	424
1951–52	162	176	17	355
1952–53	145	152	16	313
1953–54	106	130	20	274*
1954–55**	–	–	–	–

*Includes 18 seminary students not accounted for in geographical breakdown
**Bulletin/catalog not found in archives
Source: Lincoln University Archives.

(Rosenthal, 1975). Thus, another question to be considered is whether the presidents of other Black institutions were forced, as Johnson was, to place a higher priority on fund-raising and finances rather than on academic freedom, in order to ensure the financial success of their institutions. To answer this question about the two institutions in this chapter, I examined available financial reports of donations, grants, and endowments for both Cheney and Lincoln.

FUNDING, APPROPRIATIONS, AND ENDOWMENTS

Even prior to joining the Pennsylvania State System of Higher Education, both Cheyney and Lincoln received state funding for scholarships, as well as for facilities. For example, in 1951, the General State Authority (GSA) awarded contracts in the amount of $216,429 to Cheyney for the construction of a wing to Yarnall Hall, the women's dormitory (*Cheyney Record,* 1951). In addition, the GSA appropriated funds for a new power plant on the campus grounds, at a cost of $349,240. This was the seventh building constructed by the GSA since 1938 (*Cheyney Record,* 1951).

Lincoln University became a "state-aided" institution in 1938 when it received its first appropriation of $50,000 from Pennsylvania for the years from 1937 to 1939. These appropriations grew as the enrollment of Pennsylvania residents increased, but not in proportion to the enrollment numbers (*Lincoln University Bulletin,* 1951). *Lincoln's Bulletin* reported that for the years 1946–47 the institution boasted of an endowment worth over a million dollars, buildings and grounds valued at $800,000, and alumni gifts totaling approximately $350,000 for the 1946–49 period. Nonetheless, President Horace Mann Bond noted in the "Summary Factual Statement Regarding the University" that the school had been operating with "an uncomfortable deficit the past two years," and he recommended raising tuition or finding other sources of funding in addition to the state appropriations, endowed scholarships, and gifts (*Lincolnian,* January 29, 1949, p. 1). Pennsylvania Governor John S. Fine recommended to the State Board of Appropriations that Lincoln receive a $106,000 increase in the amount of funding granted for the previous two-year period. This represented a 300% increase in state appropriations for the institution between 1945 and 1953 (*LUB,* 1951, p. 13) (see Table 5.3). Lincoln sustained its financial growth because of the generation of "income from such sources as endowments, annual United Negro College Fund gifts, and state aid for scholarships" (*LUB,* 1951, p. 10). The infusion of capital assured that the institution did not have to close, as was the case with several hundred other colleges at that time (*LUB,* 1951). By the 1953–54 school year, Lincoln's endowments totaled in excess of $1 million, while buildings and grounds on the 350 acres of land were valued at $2 million. During this same period, Lincoln also received a $20,000 grant from the Ford Foundation (*LUB,* 1953–54).

Along with state and private funds, both schools also received GI tuition payments from the Veterans Administration, as well as other forms of federal assistance. As a result of war surplus properties, the government donated a new dining hall to Cheyney University in 1951. A new gym had been completed at Lincoln a few years earlier thanks to the Federal Works Agency's Veterans Educational Facilities program (*LUB,* 1946–47).

Therefore, both Cheyney and Lincoln were dependent, to some degree, upon government funding, state aid, and philanthropy. However, this alone did not support the view that they were also affected by the nationwide threat of McCarthyism. Since it would be impossible to make an assumption about the impact of McCarthyism at Black institutions without first studying actual cases where professors or administrators at either of the two universities were accused of un-American activities, of being Communists, or in which academic freedoms were being threatened, I examined school newspapers and alumni publications for reports of such cases.

Table 5.3   Record of state grants to Lincoln, 1945–1953

Years	Appropriations
1945–47	$110,000
1947–49	$154,000
1949–51	$225,000
1951–53	$331,000

Source: Lincoln University Archives.

THE MCCARTHY ERA: SUSPICION, ACCUSATIONS, AND FEAR

In the years following World War II, the threat of Communism was on the minds of most Americans. There is little doubt that the McCarthy reign exacerbated the fear that the United States was being infiltrated from within by Communists, in all walks of life (Hutcheson, 1997). In fact, reports of Communism, Communist activity, or the threat of Communism appeared routinely in student and alumni publications at both Lincoln and Cheyney. Just a year after World War II ended, a column in the November 1946 edition of *Cheyney Record* asked the question, "What is the greatest threat to world peace today?" One response was: "Communism, which is being spread to other parts of the world by Russia, is the greatest threat to a permanent world peace" (*Cheyney Record,* November 1946, p. 2). Six years later, in the same publication, a student offered insight into winning the war against Communism in a piece titled, "A World of Peace or a Realm of Fear?" (*Cheyney Record,* March 1951, p. 1). In the article the author states that "people's minds are being poisoned because of ideologies which are being thrust upon them" (p. 1), a possible reference to Communism, McCarthyism, or both.

Alumni publications also reflected the growing fear and supposed threat of Communism. In *The Cheyney University Class of 1945: Our 40th Reunion Yearbook,* listed under the "History" section for 1951, is the U.S. Supreme Court's ruling affirming that it is the state's right to require job applicants to sign non-Communist oaths. The fact that this is reported in an alumni publication clearly suggests that this was an issue at Black colleges and universities, just as it is reported to have been at White institutions (Lewis, 1988; Hutcheson, 1997; Lazarsfeld and Thielens, 1958).

In addition to the voice of alumni, students spoke out about McCarthyism in their publications. The most thought-provoking piece from a Cheyney University student appears in the May, 1951 issue of the *Cheyney Record.* Titled, "Dr. Leslie Pinckney Hill Retires: Prexy Closes Administration after Thirty-Eight Years," the article is a salute to a president who was loved and revered by the student body. A critical segment of the article reads,

> He has been smeared, denounced, and branded out of his true identity. And all of those fool hearty people who do not know the nature of their work have belittled themselves and their organization. Any man as Dr. Hill who devotes his life educating students who are to teach the very ones who are to be leaders in our democratic society needs not question his loyalty. Let the House of "No" American Activities investigate the tots who were taught by Cheyney graduates. Let them sit in on our Vespers and Chapel and Sunday School programs. Maybe, they have enough funds to visit West Chester Community Center and see what any public agent failed to do. Anyone who knows Dr. Hill would recommend him as a model citizen of America's democratic society.
>
> (*Cheyney Record,* May 1951, p. 1)

This piece appears to be a defense of Dr. Hill against a McCarthy-like assault, if not by the senator himself, then surely by some person or group enacting his policies. For therein lies the real power of McCarthyism: the methods of the man were adapted and used by so many Americans against other Americans. When questioned about the possibility that Dr. Hill had been investigated by Senator McCarthy, Beth J.

Mullaney, Cheyney University's Information Systems librarian and acting archivist, stated that she had never heard anything to that effect (personal communication, Friday, March 10, 2006).[1] However, Hill was an educator, poet, civil rights activist, and a close friend of Paul Robeson (Mullaney, personal communication, Friday, March 10, 2006), the Black singer, actor, and activist who had been aggressively investigated and interviewed by the House Committee on Un-American Activities, and who, ultimately, left America. Since it was commonplace for McCarthy investigations to extend to family members, circles of friends, and other close associates of the person under scrutiny, given the nature of the above article, the conclusion may be drawn that the unknown author implies that prior to his retirement, Hill was a victim of McCarthy-like forces.

At Lincoln, written reports also indicated students were equally aware of the threat, real or imagined, of Communism. When Ralph J. Bunche, director of the United Nation's Trusteeship Council, delivered the 1947 Commencement Address, he urged the graduates to take on the fight for civil rights, as well as world peace, with the solid support of the Constitution and the United Nations (*Lincoln Bulletin, Commencement Issue,* 1947, p. 12). However, as a reminder of the times, Bunche cautioned that "fear and suspicion among peoples and nations are universal . . . [and that] Exaggerated nationalism and imperialist ambitions constitute stubborn roadblocks in the path of true internationalism . . . [where] there is certainly no grounds for hysteria or war psychosis" (*Lincoln Bulletin, Commencement Issue,* 1947, p. 11). The selection of Bunche as their commencement speaker was a bold, as well as defiant message on the part of the Lincoln University community. As a result of his activism during the 1940s, and the fact that he had published several scholarly articles for the cause of Black equality in Communist publications (Henry, 1990), Bunche himself was accused of being a Communist. Nonetheless, on June 3, 1947, Lincoln University sent a clear message to all McCarthyites that their institution would not succumb to the political pressures of McCarthyism by asking Bunche to be the commencement speaker.

Jacques Wilmore, valedictorian of the college for the 1950 graduating class, also encouraged his peers to "stand unflinchingly for human rights, and not to give way to the name-calling hysteria of the times" (*Lincoln Bulletin, Commencement Issue,* Summer 1950, p. 3).

An even more revealing editorial was written by the student Roland V. Jones, who stated:

> The question of Communism and how great a threat it is to America is one which has come up repeatedly in small discussion groups, but it is also a question which has not come [up] enough. I say this because the Red-baiting hysteria which has swept this country has made the people afraid to discuss this question, especially if they feel there might be something said by them which would make them appear at all sympathetic toward Communism.
>
> (Jones, 1951, p. 2)

In his article Jones suggests that free thought and expression are being stifled because of McCarthyism and also questions America's failure to live up to its own democratic principles, especially in its treatment of Blacks: "The question of how great a threat

Communism is and shall be to America depends upon how great and how successful is the effort put forth by the sincere, clear-thinking Americans, Negro and White, to live up to the democratic principles set forth by the founding fathers of this nation. . . ." (p. 6). Jones offers Communism in its purist form as "the greatest form of government that this world could possibly find" (p. 2). Apparently, he was fearful of some type of repercussions, so he was cautious enough to point out that he was not a member of the Communist party himself (p. 2).

In addition to the suspicions and fear of being labeled a Communist that McCarthyism instilled on college campuses, students at Lincoln University were also battling racism and segregation at the nearby town of Oxford. They began a series of peaceful, nonviolent protests, sit-ins, and meetings with members of the Oxford community in the mid-1940s (Rosenthal, 1975). It was to be a long-drawn-out struggle that also involved charges and countercharges of Communist activity, as the towns-people sought to use the threat of McCarthyism against the activist students. In one opinion piece, a student, Stuart "Red" Dunnings, decries the failure of the American Youth for Democracy (AYD) to desegregate the city of Oxford, blaming this failure on a lack of funds (*Lincolnian,* February 26, 1947, p. 2). Originally called the Young Communist League, the AYD was the "youth wing of the Communist Party" on American college campuses. It had split up, and then reorganized in 1943 under the new name (Overstreet and Overstreet, 1958, p. 2). Comprised of both Communist and non-Communist members, the organization championed against social injustices, but was seen as a "front" for Communism (Overstreet and Overstreet, 1958). Therefore, one reason for the AYD's failure to raise funds was because of its suspected Communist affiliations. Questions about this student group had arisen at other schools, such as Wayne University in Detroit. Like Charles Johnson of Fisk, Wayne's President, David D. Henry, was Black. Henry denied any charges of subversive activity at his school and issued the following statement on the AYD:

> The University, as a public institution, requires no information from its students as to their political or religious beliefs . . . We have acted on the assumption that the University has no right to differentiate among American citizens on the basis of political beliefs insofar as admission to the University is concerned . . . We have also assumed, that if those who vote the Communist ticket or are admitted or proved supporters of Communist Doctrine are to be classified thereby as guilty of subversive activity, the legislative and police authority of the state and nation would first deal with this issue.
>
> (*Lincolnian,* May 24, 1947, p. 2)

In addition to the above theory on the AYD's funding woes, Dunnings's editorial also pointed out that the campus chapter of the National Association for the Advancement of Colored People (NAACP) experienced a similar failure a few weeks earlier, and for the same reason—lack of funds. More importantly, Dunnings questioned why the owner of the Met Theater in Oxford and the mayor of the city were both allowed to sit on the institution's board of trustees while they allowed segregation policies to be enforced (*Lincolnian,* February 26, 1947, p. 2). In order to address their funding shortage, the NAACP chapter at Lincoln "instituted a 'Freedom Fund' for the purpose of eliminating by legal prosecution, segregation and discrimination in places of public accommodation in Oxford, Pa." (*Lincolnian,* February 26, 1947, p. 1). Monies for the fund were donated by student, faculty, and alumni of Lincoln (*Lincolnian,* February 26, 1947, p. 1). Therefore, when, on January 11, 1950, four Lincoln students were arrested for sitting in the "Whites only" section of the Oxford Theatre, the Lincoln chapter of the NAACP filed a lawsuit against the owner of the theater and the two police officers involved in the arrests. The suit charged the three with "violation of federal civil right laws" (*Lincolnian,* November 1, 1952, p. 1). In response to the lawsuit, members of groups such as the Nationalist Action League made accusations of Communist activity against the NAACP. Excerpts from a letter written by W. Henry MacFarland Jr., a member of the Nationalist Action League, follow:

> As an organization dedicated to the task of containing the Red Tide of Communism at home, as well as abroad, may we respectfully request that you give the cases of Mr. Vergis and Mr. Crowl [two of the three defendants] the attention which they deserve? The record shows that the NAACP has had many affiliations with Communist front organizations, and that it has selected the peace-loving communities of Chester County for its targets in its most recently organized campaigns of agitation and unrest. Acting through its Lincoln University campus chapter, this organization has formed units of young Negroes which are dispatched to Chester County theatres, restaurants and other places of public amusement or service which maintain a policy of segregation. Once there, they proceed to seat themselves in sections reserved for white patrons, in deliberate violation of the policies in effect . . . We understand that Lincoln University is now substantially supported by funds taken from the pockets of the tax payers of this Commonwealth. This fact provides every citizen of Pennsylvania with a direct interest in the case at hand as Dr. Bond, the University's President, has led faculty in at least condoning the operations of the disturbers. We therefore intend to submit a formal demand for a complete Legislative investigation of the manner in which the University is being administered.
>
> (*Lincolnian,* February 25, 1950, p. 2)

However, there is no evidence to indicate whether or not the group followed through on its threat to request a McCarthy-like state investigation, but there is evidence in this chapter's discussion on funding and endowments to suggest that, in spite of the organization's attempt to capitalize on the anti-Communist sentiments and hysteria, Lincoln University's state appropriations were not affected.

While Lincoln's state funding may not have been diminished because of the accusation of Communist infiltration, threats of this sort could prove very damaging to the NAACP. As the major civil rights organization of the time, the NAACP's members consisted of both Blacks and Whites and the group was "committed to working through traditional mechanisms for change" (Rosenthal, 1975, p. 115). Nonetheless, unfounded allegations of Communism were often used to discredit the Civil Rights Movement and its supporters (Williamson, 2003). Such accusations, were they proven to be true, would have had a significant impact on the organization's reputation, and more importantly, its ability to raise donations from American citizens for the Civil Rights cause. Likewise, the students' affiliation with a Communist-supported organization might have proven damaging to Lincoln University's financial well-being, since the school received state appropriations and federal monies and depended upon private donations as well.

In spite of the town's use of the menace of McCarthyism, the students persevered and in 1953 a court issued a ruling on the NAACP lawsuit against Oxford. U.S. District Judge George A. Walsh "upheld the constitutionality of Pennsylvania's Non-Segregation laws," and issued two injunctions: one against all three defendants, which barred them from "interfering with the rights of *any persons* [my emphasis] to sit in any part of the theater," and a second injunction specifically against the police officers, which prohibited them "from enforcing any order of segregation of Negroes" (*Lincolnian,* October 17, 1953, p. 2). The judge also awarded $600 in damages to the plaintiffs.

The Lincoln students' success in battling racism in Oxford was seen as the "prototype for the disciplined, non-violent protests of ten and fifteen years ahead" (Rosenthal, 1975, p. 117) because they employed the sit-ins and peaceful protests that would later be adapted by the Civil Rights leaders of the 1960s. According to Rosenthal, Black veterans, whose on-campus presence is mentioned earlier in this chapter, were crucial to the success of the upcoming Civil Rights Movement, as the GI Bill of Rights increased their numbers on Black college campuses and they became more involved in campus activities. Rosenthal also acknowledges President Bond's "hands-off" policy and formal endorsement of the students' right to protest nonviolently. Bond advised his students: "Resist evil; resist it without violence. Resist evil without hatred and malice. This is the highest and hardest duty of the true Christian" (as quoted by Rosenthal, 1975, p. 118). Therefore, with the support of the veterans and their president, and in spite of the attempts by external forces to use the anti-Communist hysteria to fuel sentiments against its struggle, the NAACP campus chapter at Lincoln University was successful in its fight to desegregate the city of Oxford, Pennsylvania.

CONCLUSION

This chapter suggests that not only faculty, but administrative staff and students alike on the Cheyney and Lincoln University campuses experienced, directly and indirectly, the intimidation and fear instilled in Americans because of McCarthyism and the threat of investigations and subsequent Congressional hearings by Senator Joseph R. McCarthy and the House of Representatives' Special Committee on Un-American Activities. The results of Lloyd's survey (1952)—the Lee Lorch case at Fisk University, which Gasman (1999) details—and my own findings at both Cheyney and Lincoln universities suggest that it is likely that faculty and administrators at other Black institutions were intimidated, forced to sign oaths of allegiance, lost tenure, or were fired, just like those at White institutions. Because Black schools were more dependent upon outside sources for funds, these encroachments upon academic freedom may have been linked to funding needs, as well as political pressures put upon administrators at the Black colleges and universities. It is my hope that this chapter will generate additional studies of the effects of McCarthyism at Black institutions throughout the country, particularly at the more vulnerable schools in the South, that, like Lincoln, battled racism as well as McCarthyism. Some questions that need to be researched are: Did McCarthyism have an effect on the curriculums at Black schools? Was the threat of McCarthyism greater for Black presidents at Black schools, than for White presidents? How did Black students in the South react to McCarthyism? What were the differences, if any, between Northern and Southern McCarthyism? By addressing these and other questions, scholars will ensure the continued discourse on this important subject matter, which, until now, has been neglected by most historians of higher education.

Nᴏᴛᴇ

1. Mullaney also checked with the former archivist Eric Dulin, who admitted that, as a friend of Paul Robeson's, it is possible Hill's name was on a list of such people. However, a follow-up by Mullaney to the National Archives also proved fruitless when she received an email stating that they had "found no mention of Leslie Pinckney Hill in regard to McCarthy's hunt for Communists" (personal communication, March 10, 2006).

Rᴇꜰᴇʀᴇɴᴄᴇs

57 to receive degrees at largest commencement, vets to comprise 80 percent of grads. 1947, May. *Lincolnian* 18(7): 1–2.

American Association of University Professors. 1940 statement of principles of academic freedom and tenure with 1970 interpretative comments. Retrieved April 5, 2006, from www.aaup.org/statements/Redbook/1940stat.htm.

American Association of University Professors. 1953. Report of the ad hoc committee on freedom of teaching, research, and publication in Economics.

Bond, H. M. 1949, January. Summary factual statement regarding the university. *Lincolnian* 20(5): 1.

———. 1976. *Education for freedom: A history of Lincoln University, Pennsylvania.* Lincoln University, PA: Lincoln University Press.

Cheyney Training School for Teachers catalogue. 1947–48; 1949–50; 1950–51; 1951–52; 1952–53; 1953–54; 1954–55; 1955–57. Cheyney, PA: Cheyney University Press.

Cheyney University class of 1945: Our 40th reunion yearbook. Cheyney, PA: Cheyney University Press.

Cheyney wins new accreditation. 1951, May. *Cheyney Record* 41(5): 1.

Claptrap–a study in fascism. 1950, February. *Lincolnian* 21(5): 2.

Diamond, S. 1992. *Compromised campus: The collaboration of universities with the intelligence community, 1945 –1955.* New York: Oxford University Press.

Drewry, H. N., and D. Doermann. 2001. *Stand and prosper: Private Black colleges and their students.* Princeton, NJ: Princeton University Press.

Dr. Leslie Pinckney Hill retires: Prexy closes administration after thirty-eight years. 1951, May. *Cheyney Record* 41(5): 1, 2.

Drummey, J. J. 1987. McCarthyism, forty questions and answers about Senator Joseph McCarthy [Electronic version]. *New American* 3(10): 16–21.

Dunnings, S. 1947, February. We agree, Red! *Lincolnian* 18(5): 2.

Federal Bureau of Investigation. Freedom of information privacy act: Owen Lattimore, parts 1a and 1b. File number: 001–24628. Retrieved March 21, 2006, from http://foia.fbi.gov/foiaindex/owenlatt.htm.

Foster, H. 1948, March. Vetsville vigil. *Lincolnian* 19(5): 3.

Gasman, M. 1999. Scylla and Charybdis: Navigating the waters of academic freedom at Fisk University during Charles S. Johnson's administration (1946–1956). *American Educational Research Journal* 36(4): 739–58.

Gasman, M. 2002. W. E. B. Du Bois and Charles S. Johnson: Opposing views on philanthropic support for higher education. *History of Education Quarterly* 42(4): 493–516.

General Assembly of Pennsylvania. 2006. *House resolution no. 603.* Retrieved March 13, 2006, from http://www.legis.state.pa.us/WU01/LI/BT/2005/0/HR0603P3596.HTM.

Henry, C. P. 1990. Civil rights and national security: the case of Ralph Bunche. In *The man and his times,* ed. Benjamin Rivlin and Ralph Bunch. New York: Holmes and Meier.

Hill, S. T. 1985. *The traditionally Back institutions of higher education 1860–1982.* Washington, DC: National Center for Educational Statistics.

Holland, D. 1946, November. *Off the record. Cheyney Record* 37(1): 2.

Hutcheson, P. A. 1997. McCarthyism and the professoriate: A historiographic nightmare? In *The history of higher education,* 2nd ed., 610–27. Ed. L. F. Goodchild and H. Wechsler. ASHE Reader Series. Old Tappan, NJ: Pearson Custom Publishing.

Jones, R. V. 1951, December. Communism, America's greatest threat. *Lincolnian* 23(3): 2, 6.

Kille, J. D. 2004. *Academic freedom imperiled: The McCarthy Era at the University of Nevada.* Reno, NV: University of Nevada Press.

Lazarsfeld, P. F., and W. Thielens Jr. 1958. *The academic mind: Social scientists in a time of crisis.* Glencoe, IL: Free Press.

Lewis, Lionel. 1988. *Cold War on campus: A study of the politics of organizational control.* New Brunswick, NJ: Transaction.

Lincoln Bulletin, Commencement Issue. Summer 1947; Summer 1950. Lincoln, PA: Lincoln University Press.

Lincoln University Bulletin (LUB). 1946–47; 1948–49; 1950–51; 1953–54; 1955–56; 1956–57. Lincoln, PA: Lincoln University Press.

Lincoln University Bulletin (LUB). Winter 1951; Summer 1951; Winter 1953–54. Lincoln, PA: Lincoln University Press.

Lincoln vets sponsor second spring dance. 1947, January. *Lincolnian* 18(4): 1.

Lloyd, R. G. 1952, January. Loyalty oaths and communistic influences in Negro colleges and universities. *School and Society* 75(1933): 8.

McCormick, Charles H. 1989. *This nest of vipers: McCarthyism and higher education in the Luella Mundel Affair, 1951–52.* Urbana, IL: University of Illinois Press.

N.A.A.C.P. chapter starts freedom fund drive. 1947, February. *Lincolnian* 18(5): 1.

N.A.A.C.P. segregation suit won. 1953, October. *Lincolnian* 25(1): 2.

New wing for women's dorm. 1951, March. *Cheyney Record* 41(4): 1.

Norfolk, Virginia Journal and Guide, July 16, 1949, 1.

Overstreet, H. A., and B. W. Overstreet. 1958. *What we must know about Communism: Its beginnings, its growth, its present status.* New York: W. W. Norton.

Oxford case comes before federal court on Nov. 3, theater owner, two policemen on trial for1950 incident. 1952, November. *Lincolnian* 24(1): 1, 5.

Revlin, R., I., and P. L. Vance. 1951, March. Growing interest in Greek letter organizations. *Cheyney Record* 41(4): 1.

Rosenthal, J. 1975. Southern Black student activism: Assimilation vs. nationalism. *Journal of Negro Education* 44(2): 113–29.

Schrecker, E. 1996. *No ivory tower: McCarthyism and the universities.* Oxford: Oxford University Press.

Simkowski, J. H. 1951, March. A world of peace or a realm of fear. *Cheyney Record* 41(4): 1.

Street, H. 1958. Tactics and stratagems: The united front. *What we must know about Communism.* New York: W. W. Norton.

Thelin, J. R. 2004. *A history of American higher education.* Baltimore, MD: Johns Hopkins University Press.

Vet villagers protest removal of stoves. 1947, March. *Lincolnian* 18(6): 1.

Wayne pres. states policy on A.Y.D. 1947, May. *Lincolnian* 18(7): 2.

Williamson, J. A. 2003. Student activists, activist students: Black colleges and the civil rights movement. Paper presented at the History of Education Society Annual Meeting, Chicago, IL.

CHAPTER 6

THE FORGOTTEN GI: THE SERVICEMEN'S READJUSTMENT ACT AND BLACK COLLEGES, 1944–54

MEGHAN WILSON

One of the greatest acts of Congress, in terms of the federal government's participation in education, was the Servicemen's Readjustment Act of 1944—commonly known as the GI Bill of Rights. The bill transcended its wartime context, inextricably linking the federal government and higher education. It also afforded opportunities for all veterans to expand their educational goals, such as acquiring a college or advanced degree. Numerous scholars and historians document that the primary effect of the GI Bill was the universal increase in enrollment at colleges and universities (Adkins, 1996; Bennett, 1996; Caliver, 1945; Clark, 1998; Greenburg, 1997; Herbold, 1994–95; Hill, 1984; Hyman, 1986; Jenkins, 1946; McGrath, 1945; Olsen, 1973; Thelin, 2004; Vaile, 1944; Walters, 1944; Weaver, 1945). Enrollment increases in higher education were significant, for comparatively few industries grew as fast, gained as much prestige, or affected the lives of so many people. Enrollment at Black colleges also increased exponentially following World War II because of the GI Bill. The purpose of this chapter is twofold. First, I will illustrate how the GI Bill increased access for Black veterans to attend Black colleges. Second, I will illuminate the issues Black colleges faced because of this increase in enrollment, namely, the readjustment of the veteran student and the curricular reorganization of certain programs of study.

Many scholars from both the fields of history and education have examined the contribution of the GI Bill to postsecondary education, yet not all higher education institutions have received the same attention. Few scholars have investigated the effect of postwar enrollment of Black veterans at historically Black colleges and universities (HBCUs).

Numerous scholars have written extensively on the effects of the GI Bill on higher education during the 1940s and 1950s (Brown, 1945; Clausen and Star, 1944; Cooper, 1944; Durand, 1944; Hurd, 1945; Justice, 1946; Lindeman, 1944; McConell, 1944; Oppy and Ramseyer, 1945; Shaw, 1947). However, nearly all of the authors ignored the study of the effects of the bill on Black GIs. One researcher in particular, Martin Jenkins, departed from this trend. Jenkins conducted a series of annual surveys published in the *Journal of Negro Education* titled, "Enrollment in Institutions of Higher Education of Negroes" (Jenkins, 1945–46, 1946–47, 1947–48, 1948–49, 1949–50, 1950–51, 1951–52). Jenkins's study included every Black college and university in the nation. Most significantly, during the first four years of his survey Jenkins created a separate column listing the number of war veterans attending Black colleges. His study indicated a significant increase in enrollment of Black veterans attending Black colleges. Jenkins (1945) reported in fall 1945 that 1,310 veterans were enrolled at Black colleges. This number increased dramatically to 17,518 in fall 1946 and to 34,068 in fall 1947 (Jenkins, 1946, p. 229; Jenkins, 1947, p. 211). Jenkins's study is essential for meticulously documenting the increase in veterans at Black colleges post World War II.

According to Hill (1984), in 1942 the federal government published the *National Survey of the Higher Education of Negroes*, which examined the progress and problems of the Black colleges within a social, economic, educational, and political context. This survey provided me with a look into the various issues affecting Black colleges before the creation of the GI Bill. Information on enrollment, tuition, curriculum, and faculty was important in terms of understanding the unique context before the war's end.

While the majority of research on the postwar enrollment of Black veterans at Black colleges was written over 50 years ago, two contemporary scholars have examined various aspects of this topic (Onkst, 1998; Turner and Bound, 2003). For instance, David Onkst (1998) studied a particular segment of Black Americans by analyzing what happened to Black veterans in the Deep South when they tried to obtain their GI entitlements. In the article, "'First a Negro . . . Incidentally a Veteran': Black World War Two Veterans and the GI Bill of Rights in the Deep South, 1944–1948," Onkst (1998) found that through a combination of racial discrimination and poor administration of the bill's benefits, Black veterans in the Deep South struggled tremendously attempting to go to college. With regard to education he stated, "Black colleges in the South only admitted veterans who had completed high school, or who could pass a high school equivalency test. For many Black veterans, such admission standards became a major obstacle" (Onkst, 1998, p. 527). For those who did enroll in postsecondary education, the conditions proved to be deplorable: "Old dilapidated buildings with virtually no doors or window, a very poor heating system and most despairing of all poor teachers and worse subject matter" (Bohannon in Onkst, 1998, p. 528).

In a second article, "Closing the Gap or Widening the Divide: The Effects of the GI Bill and World War II on the Educational Outcomes of Black Americans," Turner and Bound (2003) use empirical evidence to determine whether the GI Bill and World War II had a positive effect on the educational attainment of Black Americans. The central question of their analysis is "whether the effect on collegiate attainment of military participation and the availability of educational benefits varied with race and geographic location" (Turner and Bound, 2003, p. 154). Ultimately, Turner and Bound (2003, p. 145) found that "World War II and the availability of GI benefits had a substantial and positive impact on the educational attainment of White men and Black men born outside the South. However, for those black veterans likely to be limited to the South in their educational choices, the GI Bill had little effect on collegiate outcomes." Similar to Onkst's (1998), Turner and Bound's (2003, p. 153) study concluded that "historically black colleges were more limited than White colleges in their ability to accommodate returning servicemen because institutional resources were more scarce and deficiencies in physical space were often more serious than white institutions." Of note, however, only a small portion of each study focuses exclusively on Black colleges and universities. Regardless, both articles contribute to the limited existing literature available on the Black veteran experience. More importantly, they each serve as a catalyst to reawaken a topic that has remained untouched for many years.

THE GI BILL AND HIGHER EDUCATION

The most significant change as a result of the GI Bill was the "increased access to higher education for a vast majority of a population where college was not a reality" (Thelin, 2004, p. 184). Immediate postwar enrollment figures approximate a universal increase of 1,750,000 students to colleges and universities (Vaile, 1944, p. 53). Higher education opportunities opened enrollment to more groups than in the years past. Education served as a social safety valve that eased the traumas and tensions of adjustment from wartime to peace.

Black colleges also grew substantially in the three decades prior to World War II. Hill (1984) reports the growth in both private and public Black colleges from 1914 to 1954. Private Black college enrollment increased from 20,138 students during the academic year 1935–36 to 30,772 during the academic year 1953–54. At the same time, public Black colleges enrolled 13,065 students during the academic year 1935–36 and 43,119 students during the academic year 1953–54. Veterans represented a fairly large proportion of the overall growth in enrollment at Black colleges following World War II. Table 6.1 depicts the number of veterans enrolled in both private and public Black

Table 6.1   Veterans enrolled in Black colleges (private and public), 1946–1952

Fall of year	Number of veterans	Total students	Percentage of all students
1946	18,216	58,842	31
1947	26,306	74,173	35
1948	22,526	70,644	32
1949	19,320	70,431	27
1950	13,562	69,651	19
1951	7,985	66,290	12
1952	4,222	68,375	9

Source: Hill, 1984, *Traditionally Black Institutions*, p. 13.

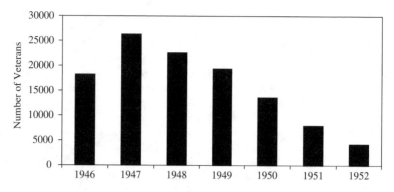

Source: Hill, 1984, *Traditionally Black Institutions*, p. 13.

Figure 6.1   Veteran enrollment in HBCUs from 1946 to 1952.

colleges in the Southern states from 1946 to 1952. Figure 6.1 visually represents total veteran enrollment at public and private Black colleges. In 1947, veterans made up over one-third of the total enrollment at Black colleges, a percentage that remained significant into the early 1950s.

Jenkins, a graduate and scholar from Howard University, conducted a series of studies examining African American enrollment at Black colleges following World War II. Over the course of eight years (1945–53) Jenkins surveyed approximately 112 institutions in 20 states plus the District of Columbia in order to determine enrollment patterns and practices of Black colleges. During the first four years of his study, 1945–49, he calculated the number of veterans at *each* institution. For the remaining three years, 1950–52, Jenkins calculated the total enrollment for veterans at *all* institutions combined.

One can note the considerable increase in enrollment during the first three years of Jenkins's survey. In fall 1945, Jenkins reported, "1,310 veterans took work on the college level in 85 institutions" (Jenkins, 1946, p. 236). One year later, the total number of veterans attending Black colleges increased 13-fold to 21,767, and constituted approximately 30% of the total enrollment at Black colleges (Jenkins, 1947). Veteran enrollment reached an all time high in fall 1947 to just over 29,500 students. Of this number, the majority (81%) of students were undergraduate; however, some students (15%) were taking classes below college level. Total enrollment reached its peak in 1947–48 and gradually declined over the next few years as enrolled veterans completed their coursework. The following year, in fall 1948, Jenkins (1949) noted a decrease of approximately 25% in veteran enrollment. Even though the total number of veterans dropped a bit from the previous year (1947), the total number of veterans in fall 1948 still exceeded the total number of veterans in fall 1945 by 16,792, or by almost 400%.

For the remaining three years that Jenkins conducted his study (fall, 1949; fall, 1950; and fall, 1951), the number of Black veterans attending Black colleges continued to decline. In 1949, Jenkins reported, "Eighty-five institutions report a total of 14,303 veterans enrolled during the fall term in undergraduate courses." In 1950, 96 institutions reported a total of 11,033 veterans enrolled during the fall term in undergraduate courses, a 28% decrease from the previous year (Jenkins, 1950, p. 216). In 1951, only 5,807 veterans were enrolled in undergraduate courses, a 41% decrease from the previous year (Jenkins, 1952, p. 214). The initial wave of Black veterans who received their GI Bill benefits graduated by the late 1940s. Jenkins (1950) speculated that although undergraduate enrollment would continue to decline, graduate and professional enrollment would increase as students continued life in the academy.

In the initial publication of the results of his survey, Jenkins (1946) urged Black colleges to factor the steady increase in enrollment following the war into the planning of educational programs. Jenkins (1946, p. 239) added, "The writer is convinced that the major implication of increasing enrollment is the need for the institutions to find an answer to the problem of providing an adequate educational environment for their students." Creating such an "environment" would prove especially difficult. The next section of the chapter explores the challenges Black colleges faced because of the overall increases in the size of their student bodies, namely, overcrowded conditions, staffing shortages, and inadequate funding.

THE CONSEQUENCES OF HIGHER ENROLLMENT

The need for additional facilities at Black colleges was severe (Atkins, 1948; Turner and Bound, 2003). The great majority of these colleges were located in the poorest section of the country, namely the Deep South (Onkst, 1998). Financial support for education in that region was less than in any other part of the country (Atkins, 1948). Black colleges repeatedly suffered from a lack of equitable division of resources, despite the aforementioned upward trends in enrollment. Across all private and public institutions in the South, "White institutions accounted for 92% of total expenditures in 1943–45; among public institutions alone, colleges and universities for Whites accounted for more than 94% of [federal] expenditures" (Turner and Bound, 2003, p. 152). According to Public Law 697, an amendment to Title V of the Lanham Act, White institutions received the majority of the allocations (Public Law 849, 76th Congress, approved October 14, 1940). Congress passed the law in order "to relieve acute shortages of educational facilities required for persons engaged in the pursuit of courses of training or education under Title II of the Servicemen's Readjustment Act of 1944" (Sect. 504, 1940). As a result, historically White institutions (HWI) secured nearly triple the amount of allocations compared to their Black counterparts (Atkins, 1948). In 1946 and 1947, veterans constituted a larger percentage of total enrollments at White institutions than at Black colleges. However, the percentage increase in the number of veterans at White institutions in 1947 was 29.4% as compared with 50.0% at the Black institutions (Office of Education, Circular No. 238, 1947). Therefore, Black colleges and universities experienced greater enrollment pressure from the servicemen eligible to receive benefits from the GI Bill than HWIs. Overall, Black institutions received only about "40% of what the colleges estimated that they would require" to provide academic programs (Atkins, 1948, p. 152). The deficiencies in fiscal resources and physical space limited Black colleges' ability to accommodate returning servicemen.

The inability of Black colleges to effectively digest increased student enrollment resulted in a struggle for adequate space in both the classroom and the dorm room. Jenkins (1946, p. 239) described "English composition classes of 80 students and elementary language classes of 60 students taught by a single teacher, without assistants." Students often sat on windowsills or the floor because desks were a precious commodity. On-campus housing presented additional challenges, a setback for colleges and universities where "six students [were] sleeping and studying in a dormitory room designed for two" at some schools (Jenkins, 1946, p. 239).

THE BLACK GI EXPERIENCE

The GI experience for Black Americans differed markedly from their White counterparts during and after World War II as was reflective of America's racial dichotomy. American soldiers fought under a system of legal racial segregation (Motley, 1975). During the war, Black soldiers and sailors struggled with their own government to gain equal access to military positions long before they fought against the Axis enemy (Motley, 1975). The military started to open its doors to Blacks because of the enactment of the Selective Service Act of 1940, which prohibited racial discrimination in military recruitment (Congress of United States, 54 Stat. 885). "The new policy required the armed forces to allow the enlistment of Blacks according to their proportion of the population" (Weaver, 1945). President Roosevelt's policy maintained segregated units, but made Blacks eligible to serve in the Army Air Corps, eligible for officer training, and eligible for civilian jobs within the military (Weaver, 1945).

With the changes of 1940, the army accepted Black troops in greater numbers; nevertheless, Blacks were still rejected from service more frequently than Whites, and they were relegated to service units rather than to combat units more often than Whites even toward the end of the war (Turner and Bound, 2003). Most Black soldiers who entered the military were poor and relatively uneducated (Trudeau, 1945; Ware and Determan, 1966). At first the Selective Service did not regard this as a great problem, but when the demands of the war became critical and it realized that the armed forces could lose the services of 750,000 Black civilians because they were illiterate, changes were made. By 1944 more than one million Blacks were already inducted into the armed forces of the United States; however, these soldiers were untrained and uneducated because of both the educational and occupational limitations in the country prior to World War II (Weaver, 1945; Schiffman, 1949). For Blacks in the armed forces this meant, along with Whites, schooling in mechanical skills for the war. The ability to enlist in the armed forces was a long-sought opportunity for Black civilians and hundreds of thousands eagerly pursued it.

According to the 1940 census, of the approximate three million employed Blacks, 42.2% were employed at farms (United States Census Bureau, 1940). Therefore, at the time of their induction to the armed forces a majority of Blacks were engaged in agricultural work. Black civilians received little experience or opportunity for training or employment in semiskilled and skilled work. Training for the mechanized war proved to have significant influences on the work characteristics of inductees. Weaver (1945, p. 134) took notice of the importance of the training for Black soldiers: "New fields of activity have been opened to a sizable number of colored men. Many of these occupations—machine work, radar, radio, refrigeration, electricity, sheet metal—are the very types of work from which Negroes have been constantly barred in the past." Wartime training prepared the Black GI for participation in new occupations and industries. Tens of thousands of Black veterans formerly engaged in agriculture could now attempt to enter industrial, commercial, and service employment after the war.

POSTWAR EDUCATION

Following World War II the Black solider had two alternatives: to pursue gainful employment or to continue his education by enrolling in a college and university. Weaver (1945, p. 130) reported,

> The returning Negro serviceman will come back to an economy in which the position of the Negro has changed. He will come back to an economy in which colored workers have been employed in a more diversified occupational, industrial, and individual establishment pattern. He will come back to an economy in which his civilian relatives and friends have tasted of higher-skilled and better paid jobs. The ethnic group to which he belongs will be aggressively demanding better job opportunities and the retention of recent economic gains. The Negro veteran will be prepared to participate in new occupations and industries; he will press for such participation.

President Truman supported an alternative pathway for the Black veteran; namely, attending college. In a letter acknowledging the fourth annual drive of the United Negro College Fund (UNCF), he wrote,

> We have a great obligation to the hundreds of thousands of Negro veterans who served in the war in all capacities and in all services. It is right that we should discharge a part of this obligation by extending and improving the facilities for higher education among Negroes. That is the American way. It has not failed in the past; it will not fail us now.
>
> (cited in the *New York Times*, April 8, 1947, p. 24)

Yet the "American way" referenced by President Truman did not necessarily translate into immediate access to higher education for Black GIs after World War II. Black servicemen experienced difficulty entering college because of overcrowded campuses or inadequate preparation for college-level work (Turner and Bound, 2003). In addition, most White universities "discouraged blacks from matriculating, and official or unofficial quotas existed at those places that did admit blacks" (Herbold, 1994–95, p. 107). Because of overcrowding and impossible admission standards at predominately White institutions, many Blacks were forced into vocational training programs or trade schools. On the other hand, one group of institutions opened their doors to Black veterans—Black colleges.

THE RESPONSIBILITY OF BLACK COLLEGES TO THE BLACK GI

During the postwar years, numerous Black scholars argued that Negro colleges had the responsibility to offer an education to Black veterans (Caliver, 1945; Trudeau,

1945; Jenkins, 1944, 1946; Fitchett, 1945). According to Schiffman (1949, p. 23), as educational leaders, Black colleges could "do much to capitalize on this educational interest expressed by Negroes." Fitchett (1945, p. 94) added,

> It is fair to say that the responsibility of an educational institution is contingent upon problems incident to the ways of life which the social system imposes upon its members. When these problems thwart the efforts of a group to function effectively towards the realization of the cherished treads, principles and ideals of the social system, it is incumbent upon the school to operate in those spheres of difficulty.

In his bold declaration, Fitchett held educational institutions directly accountable to not only function within a difficult environment but also to supersede that environment and challenge the current social system. Black colleges, in particular, were educational institutions uniquely positioned to work within these "spheres of difficulty," and offer access to higher education for returning Black servicemen.

READJUSTMENT

At the end of the war, upward of 100,000 Blacks in the army had the equivalent of a high school education or better (Trudeau, 1945). Many planned to continue with further schooling as a result of the benefits provided by the GI Bill (Wright, 1947). As educational leaders, Black institutions needed to capitalize on this educational interest, if not from a social standpoint then from a financial one. Veterans' tuition to postsecondary institutions was greatly subsided by the federal government, as part of the soldier's benefits package (McGrath, 1945). Secured tuition helped Black colleges grow and prosper (Ware and Determan, 1966). Before classes began, Black colleges first needed to understand the particular needs of returning servicemen. In the article, "The Background of the 1947 College Student," Benjamin Quarles (1947) provided characteristics for Black colleges about the current student body, namely, the Black GI. For the 1947 fall semester, he explained that the average student is male and older than the typified freshman of 1941. Additionally, the increased veteran enrollment to Black colleges brought a larger percent of students whose parents belong to lower occupational levels and a higher proportion of students coming from financially impoverished and culturally limited homes (Quarles, 1947). Finally, Quarles (1947) noted the trend toward vocational education. Shaw (1947, p. 18) wrote, "A recent study shows that 82% of prospective student veterans want courses that are functional and utilitarian." Quarles (1947, p. 89) agreed and stated, "The Negro veteran's interest in vocation is at a high pitch."

Fitchett (1945) examined the ramifications of the large numbers of Black veterans enrolling at Black colleges. He assumed that returning veterans might experience challenges adjusting to the life of a college student; so he designed a survey that he sent to Black colleges, Black educators, and Black soldiers. In the survey, Fitchett (1945) asked each respondent what he or she considered the most difficult area of a veteran's readjustment back to civilian life. Amongst the most popular responses were "health—physical and mental, civic and political associations and vocational and economic concerns" (p. 96). Jenkins (1946, p. 239) also made a plea to Black colleges to appreciate the unique needs of the Black GI. He wrote, "The service of a college to its clientele is to be estimated not so much in the *number* of students it has, but rather in *what it does to and for the students it has.*" Black colleges needed not only to educate but also to rehabilitate their returning students.

Before considering ways to reformulate their curricula to meet the needs of their new population, Black colleges and universities needed to ensure the successful readjustment of veterans into civilian life. Wright (1947, p. 246) defined adjustment as "the process of finding and adopting modes of behavior suitable to a given environment and this environment may be a college, the larger community, or the democratic way of life, or preferably all these." The findings from the *Postwar Education of Negroes* (Caliver, 1945, p. 1) concluded that "Negro veterans and war workers needed assistance in readjusting to normal peacetime living, first because of the changes which have taken place in them, and second because of the changes that have taken place in their home communities." Because so much of the Black veteran's adjustment needed comprised matters relating to education, and because every community had one or more educational institutions, schools and colleges were among the most important agencies, which had a responsibility of initiating, prosecuting, and coordinating plans and activities for the Black veteran's adjustments in the postwar period. Educational institutions, especially Black colleges, were challenged with overdue curricular revisions necessary to keep pace with the changing demands of the labor market (Quarles, 1947; Weaver, 1945). Caliver (1945, p. 3) explained some of the curricular revisions Black colleges needed to meet the general population in the postwar period:

> Revision of educational objectives; improvement in evaluating educational achievement; adaptation of curriculum to needs; provision of more and better qualified counselors and teachers of adults; and improvement in instruction through personnel studies, acceleration, and new materials and aids. In addition to preparing to assist the veteran and the war worker to make the necessary adjustments, education will be called upon to revise its concept and program on all levels to meet the needs of the general population in a new age.

The need for these overdue changes had been emphasized by the war and was increasingly accentuated by the postwar situation. To keep pace with the new demands in the labor market colleges and universities needed to reconceptualize the way they organized their curriculum. Granger (1945) declared, "I hope that Negro colleges will prepare for their full responsibility in post-war American-life by placing greater emphasis on instruction and guidance of students in other than formal classroom programs" (Letter, February 28, 1945). Thus, a curricular overhaul was necessary to meet the needs of the veteran student, to meet labor market demands, and to help veterans readjust back to civilian life.

CONCLUSION

Many Black veterans were optimistic about the possibilities of continuing their education, despite some disbelief shared among some Americans. In fact, the proportion of Black enlisted men who expressed a desire for education after the war exceeded that of White enlisted men (Trudeau, 1945). The numbers reveal the magnitude of educational interest expressed by Black veterans. Herbold (1994–95, p. 106) reported that "over ten years postwar enrollments at black colleges increased almost 100% from 43,003 to 76,600." By the mid-1940s approximately "3,000 Blacks held master's degrees and more than 550 earned Ph.D.s" (Anderson, 1993, p. 157).

As the result of increased access and enrollment, Black colleges shifted their focus to meet the needs of their new study body: the Black GI. Both of these goals, *readjustment* and *a reorganization of the present curricula*, were the responsibility of the Black institutions. Black colleges aided veterans' *readjustment* back into society since many Black veterans were awarded opportunities that were previously not provided to them (Wright, 1947; Schiffman, 1949). Rather than predominately working in agriculture, Blacks could attend college and secure a job in any number of fields (Weaver, 1945). The Veterans Administration was typically responsible for assisting with veteran's transition back to peacetime life; however, because of the inconsistencies in interpretation and implementation of war benefits for Black soldiers, the responsibility fell to the Black colleges.

By accepting readjustment as a primary goal, Black colleges *reorganized* their curricular structure. The prewar trend to offer a more rounded, less specialized type of education was obsolete (McGrath, 1945). Many students who passed through the specialized training schools of the army and the navy received instruction in scientific fields, such as physics, chemistry, and biology (McGrath, 1945). As a result of their training experiences in the army, "many Negroes [were] more attentive, in the postwar period, to the possibilities of education" (Trudeau, 1945, pg. 89). Prior training, coupled with student interest and attention, required Black colleges to shift their curricular focus to best meet the needs of their new student body.

This investigation serves as a starting point for future research concerning Black veterans and Black colleges. What is adequately documented is the responsibility Black colleges accepted to help educate and rehabilitate returning veterans. Additionally, evidence is provided that illustrates increased enrollment at Black colleges as a result of the GI Bill. And lastly, further research is needed to document some of the changes implemented by Black colleges as a result of higher enrollments of veterans.

REFERENCES

Adkins, W. 1996. Changing images: The GI Bill, the colleges, and American ideology. *Journal of the Georgia Association of Historians* 17: 18–40.

Anderson, J. 1993. Race, meritocracy, and the American academy during the immediate post-World War II era. *History of Education Quarterly* 33(2): 151–75.

Atkins, J. 1948. Negro educational institutions and the veterans' educational facilities. *Journal of Negro Education* 17(2): 141–53.

Bennett, M. J. 1996. *When dreams came true: The GI Bill and the making of modern America.* Washington, DC: Brassey's.

Brown, F. J. 1945. Schools and colleges serve the disabled. *Annals of the American Academy of Political and Social Science* 239: 165–74.

Caliver, A. 1945. *Postwar education of Negroes.* Washington, DC: United States Office of Education.

Clark, D. A. 1998. "Two Joes meet–Joe college, Joe veteran": The GI Bill, college education, and postwar American culture. *History of Education Quarterly* 38(2): 165–89.

Clausen, J., and S. Star. 1944. The soldier looks ahead. *Annals of the American Academy of Political and Social Science* 231: 9–17.

Cooper, W. 1944. Reorganizing and adapting the college curriculum to meet the post-War needs of soldiers, war workers, and youth in general. *Quarterly Review of Higher Education among Negroes* 12(2): 85–90.

Durand, J. 1944. Veterans and war workers in the postwar labor force. *Annals of the American Academy of Political and Social Science* 231: 27–36.

Fitchett, H. E. 1945. What are the special responsibilities of Negro colleges for the adult education of negroes in the post-war world? *Quarterly Review of Higher Education among Negroes* 12(2): 94–110.

Granger, M. February 28, 1945. Personal letter.

Greenburg, M. 1997. *The GI Bill: The law that changed America*. West Palm Beach, FL: Lickle Publishing.

Herbold, H. Winter, 1994–95. Never a level playing field: Blacks and the GI Bill. *Journal of Blacks in Higher Education* 6: 104–8.

Hill, S. 1984. *The traditionally Black institutions of higher education 1860–1982*. Washington, DC: National Center for Education Statistics.

Hurd, C. 1945. Educational plans questioned as lacking in realistic touch. *New York Times*, February 18, 28.

———. 1946. 300,000 veterans are taking courses today in universities and colleges, with 162,485 vacancies expected in fall. *New York Times*, May 5, 39.

Hyman, H. M. 1986. *American singularity: The 1787 Northwest Ordinance, the 1862 Homestead and Morrill Acts, and the 1944 GI Bill*. Athens, GA: University of Georgia Press.

Jenkins, M. 1946. Enrollment in institutions of higher education for Negroes, 1945–46. *Journal of Negro Education* 15(2): 231–9.

———. 1947. Enrollment in institutions of higher education for Negroes, 1946–47. *Journal of Negro Education* 16(2): 224–32.

———. 1948. Enrollment in institutions of higher education for Negroes, 1947–48. *Journal of Negro Education* 17(2): 206–15.

———. 1949. Enrollment in institutions of higher education for Negroes, 1948–49. *Journal of Negro Education* 18(4): 568–75.

———. 1950. Enrollment in institutions of higher education for Negroes, 1949–50. *Journal of Negro Education* 19(2): 197–208.

———. 1951. Enrollment in institutions of higher education for Negroes, 1950–51. *Journal of Negro Education* 20(2): 207–22.

———. 1952. Enrollment in institutions of higher education for Negroes, 1951–52. *Journal of Negro Education* 21(2): 205–19.

Johnson, H. 1947. The Negro veteran fights for freedom. *Political Affairs*, 430.

Justice, T. G. April, 1946. What happens to the veteran in college? *Journal of Higher Education* 17(4): 85–188; 224–5.

Lindeman, E. 1944. New needs for adult education. *Annals of the American Academy of Political and Social Science* 231: 115–22.

McConnell, T. R. 1944. Liberal education after the war. *Annals of the American Academy of Political and Social Science* 231: 81–7.

McGrath, E. J. March, 1945. Postwar jobs for veterans. *Annals of the American Academy of Political and Social Science* 238: 77–88.

Motley, M.P. 1975. *The invisible soldier: The experience of the Black soldier, World War II.* Detroit, MI: Wayne State University Press.

Office of Education. 1947. Circular No. 238. Washington, DC.

Olsen, K. 1973. The GI Bill and higher education: A success and surprise. *American Quarterly* 25(5): 596–610.

Onkst, D. 1998. "First a Negro . . . incidentally a veteran": Black World War Two veterans and the GI Bill of Rights in the Deep South, 1944–1948. *Journal of Social History* 31(3): 517–43.

Oppy, G., and J. Ramseyer. 1945. *Secondary education for veterans of World War II.* Columbus, OH: Ohio State University Press.

Quarles, B. 1947. The background of the 1947 college student. *Quarterly Review of Higher Education among Negroes* 15(2): 87–90.

Schiffman, J. 1949. The education of Negro soldiers in World War II. *Journal of Negro Education* 18(1): 22–28.

Shaw, R. M. January, 1947. The GI challenge to the colleges. *Journal of Higher Education* 18(1): 18–21.

Thelin, J. R. 2004. *A history of American higher education.* Baltimore, MD: Johns Hopkins University Press.

Trudeau, A. 1945. The role of Negro schools in the post-war period. *Quarterly Review of Higher Education among Negroes* 13(4): 84–93.

Turner, S., and J. Bound. 2003. Closing the gap or widening the divide: The effects of the GI Bill and World War II on the educational outcomes of Black Americans. *Journal of Economic History* 63(1): 145–77.

United States Census Bureau. 1940. Washington, DC.

United States Veterans Administration. 1950. *Report on education and training under the Servicemen's readjustment act.* Washington, DC: U.S. Government Printing Office.

Vaile, R. January, 1944. Enrollment after the war. *Annals of the American Academy of Political and Social Science* 231: 53–7.

Walters, R. 1944. Facts and figures of colleges at war. *Annals of the American Academy of Political and Social Science* 231: 8–13.

Ware, G., and D. W. Determan. 1966. The federal dollar, the Negro college, and the Negro student. *Journal of Negro Education* 35(4): 459–68.

Weaver, R. 1945. The Negro veteran. *Annals of the American Academy of Political and Social Science* 238: 127–32.

Wright, S. 1947. The admission, counseling, adjustment, and achievement of veterans. *Quarterly Review of Higher Education among Negroes* 15(4): 243–50.

CHAPTER 7

RACE, SOCIAL JUSTICE, AND THE JACKSON STATE UNIVERSITY SHOOTINGS

MARK S. GILES

In American higher education, race and class matter. The origins, functions, and missions of historically Black colleges and universities (HBCUs) starkly illustrate this point. These institutions were born out of unique struggles for social justice and the American political and funding context of pacification toward those African Americans. HBCUs were created as institutions to uplift and educate Blacks and as a means to support the national policy of apartheid between Black and White learners.

America's checkered history in providing equal and fair educational opportunities for all citizens is at worst disgraceful, and at best unequal. The myths of meritocracy combined with norms of elitism, racism, sexism, and economic bias form the climate and culture of what marginalized groups have endured and overcome. The policy and practice of White privilege formed the systemic landscape of historically White institutions (HWIs) and influential public perceptions of who deserves or does not deserve certain levels of treatment (Wise, 2005). The historical record of American higher education can provide a lens to review issues of social justice and the critical roles of race, class, gender, and power.

Black colleges and universities, because of the inherent nature of their existence in an American context, present social justice models that opened the doors of educational opportunity to marginalized and oppressed groups that, until the mid-1950s, were largely denied access to higher education. Of course, many HBCUs were founded in the late nineteenth and early twentieth centuries; nevertheless, it took social struggle in the form of court cases, civil rights movements, and public outcry to change the majority of HWIs from a mostly biased exclusive set of educational institutions to ones that are still struggling with full inclusion. The role of HBCUs in educating Black Americans is complex and filled with triumphs and tragedies (Williams and Ashley, 2004). This chapter highlights one of the dark moments in that history.

On May 14, 1970, in Jackson, Mississippi, on the campus of Jackson State College, a Black college, a mere ten days after the infamous Kent State shootings, Jackson police and Mississippi Highway patrolmen repeatedly fired their weapons at and into a women's dormitory. Two unarmed Black men were killed. One victim was a Jackson State student sitting in front of the women's hall and the other victim was a local high school student who happened to be walking across the campus on his way home from an after-school job. According to official accounts, 12 Jackson State students were wounded by the gunfire and received treatment at a local hospital or through private physicians (Spofford, 1988). Although it was one of only several campus shootings of college students by law enforcement agents during the late 1960s and early 1970s, it is often overlooked by historians of higher education. Why is the Jackson State shooting an often overlooked event in the popular narratives of the turbulent 1960s in higher education history? It received less attention at the time than the more well-known Kent State shootings ten days earlier (Dionne, 1990), and it continues to garner less attention in 2007 when discussions of the history of Black colleges occur.

CRITICAL RACE THEORY

It should be remembered that those shootings did not take place in a vacuum, nor were they the first or last to occur on a college campus. In this chapter, I contend that the racial, social, and political atmosphere of America in 1970, as well as the unique culture of the South, created a "perfect storm" environment for the Jackson State incident to occur and for it to become buried within the mainstream national memory. Insights from critical race theory (CRT) help frame my position and the incident. Richard Delgado (1995, p. xiv) states, ". . . racism is normal not aberrant, in American society. Because racism is an ingrained feature of our landscape, it looks ordinary and natural to persons in the culture." This analysis provides one lens through which to view why the killings and surrounding events did not receive widespread coverage and why it is not more well known today. Obvious facts are offered for consideration. The institution was and is predominately Black. The victims, dead and wounded, were Black. The shooters (i.e., Mississippi state troopers) were White. Although several civil and criminal cases have been brought before the courts, all have been dismissed without finding in favor of damages for victims, or to hold anyone criminally liable (Lesher, 1971; Dionne, 1990). *New York Times* journalist Stephan Lesher, writing one year after the shooting proved prophetic, noted, "No one has been punished. No one is going to be. There is the deep-seated and historically justified belief among Jackson's Blacks that but for the proximity in time to the killing of four white students at Kent State University, the Jackson State shootings would have attracted scant national attention" (p. SM24). Another important consideration in using CRT as a lens is to keep in focus the context in which the issue or event is examined.

According to Ladson-Billings and Tate (1995, p. 57), "critical race theorists argue that political and moral analysis is situational . . . For the critical race theorist, social reality is constructed by the formulation and the exchange of stories about individual situations. These stories serve as interpretive structures by which we impose order on experience and it on us." Crenshaw et al. (1995, p. xiii) assert,

> Although Critical Race scholarship differs in object, argument, accent, and emphasis, it is nonetheless unified by two common interests . . . [first] to understand how a regime of white supremacy and its subordination of people of color have been created and maintained in America, and, in particular, to examine the relationship between that social structure and professed ideals such as "the rule of law" and "equal protection" . . . [second] a desire not merely to understand the vexed bond between law and racial power but to change it.

Understanding what happened at Jackson State and why it happened is impotent without maintaining a clear focus on the context of race and racism in the South and in America during the latter stages of the Civil Rights Movement, the political climate of "law and order," and the years of Vietnam War escalation. This chapter highlights the "voices" of several survivors of the shooting.

1970: A YEAR OF VIOLENCE AGAINST COLLEGE STUDENTS

Kenneth W. Goings, associate professor and chairman, Department of History, Rhodes College, Memphis, Tennessee, was a student at Kent State in 1970 when those shootings occurred, and has written extensively on the subject. Goings (1990, p. 58) writes,

> [The] Kent State killings were not the first nor the last of that era. Tragedies involving Black students at South Carolina State College and Jackson State College received little press and remain largely unknown to the American public. This disregard was not just an isolated incident. It was typical of the unequal treatment of African-Americans and White Americans in the press and by historians.

Jackson State was not the only HBCU where Black students were killed by law enforcement officers during this era. Rosenthal (1975, p. 128) reports,

> Three students were shot to death by police at South Carolina State College in February 1968. An honor student was killed during a shoot-out between students and police in May, 1969, at North Carolina A&T in Greensboro . . . Finally, two students were killed when sheriff's deputies re-took the Southern University administration building in Baton Rouge, Louisiana on November 16, 1972.

In researching this topic, I found many passing references to the Jackson State shootings in a variety of sources, but very little in-depth literature specifically about the incident. Historian Christopher Lucas (1994) devotes only two paragraphs to both the Kent State and Jackson State shootings out of a 316-page history of American higher education. John Thelin (2004) mentions Jackson State in one paragraph that also mentions the shootings at Kent State. The most useful information I found was from articles written by eyewitnesses, Young (1990), and Weakley (1990); reporters Coombs (1973) and Huge (1970); and books by O'Neil, Morris, and Mack (1972), Rhodes (1979), and Spofford (1988). In 1972, Robert O'Neil, John P. Morris, and Raymond Mack published *No Heroes, No Villains*. This is a thoughtful comparative work of the

Kent State and Jackson State shootings. Though informative, it lacks sufficient historical perspective because it was written immediately after the incidents.

Spofford's *Lynch Street* is perhaps the most definitive account of the Jackson State shootings and its immediate aftermath. It also is the source that offers the historical perspective missing from earlier books and reports. According to Spofford (1988), a journalist with the *Albany Times-Union* at the time of the book's publication, he reviewed FBI files, combed through local and national newspapers, used archival and secondary sources, and conducted 120 interviews. Thus, he crafted a well-documented work that extends the first-hand accounts and reports offered in other published works. I relied heavily upon *Lynch Street* to inform my understanding of what took place at Jackson State in May 1970. Clearly there are limitations for relying too heavily upon secondary sources for historical research, yet it is a useful method for understanding context and considering additional resources that might shed multiple perspectives upon the particular subject under investigation. Before examining the actual events leading up to and including the shootings, a brief look at Jackson State's history offers an interesting view of how and why it was founded.

JACKSON STATE COLLEGE: SERVING THE NEED FOR NEGRO HIGHER EDUCATION

Jackson State is an HBCU. It originally began as Natchez Seminary, in Natchez, Mississippi, in 1877, the same year Reconstruction ended (Rhodes, 1979; Coombs, 1973). Similar to other colleges for Blacks founded during the latter part of the nineteenth century, the political maneuverings, motives of key players, and financial woes convey the complexity of starting an institution of advanced learning for former slaves and their descendants. Soon after the Civil War, the need to educate the freedmen became a social and economic reality in the South; however, it did not happen without resistance or controversy. Rhodes (1979, p. 5) states,

> [B]y the end of Reconstruction, Southern Whites began to realize the necessity of educating the freedmen to fit them into their new social life. On the other hand, some whites continued to oppose any kind of education for the freedmen. The numerous fears expressed included the thought of government interference, destruction of the southern caste system, and, more subtly, fear of miscegenation.

The Freedman's Bureau in Washington, DC, along with the American Baptist Home Missionary Society, played key roles in the founding of Jackson State College. More importantly, it was the leadership of Black Baptist ministers in Mississippi that gave birth to the idea of an institution to educate Mississippi's freedmen (Rhodes, 1979). This example of agency and valuing education through religious networks helps to displace the myth of Black passiveness and dependence on Whites for education and uplift. Specifically, the genesis of Natchez Seminary (Jackson State College) originated from a group of Black Baptist ministers who attended the first Saint's Baptist Missionary Association convention held at King Solomon Baptist Church in Vicksburg, Mississippi, in 1869 (Rhodes, 1979). A central figure at that meeting and at the forefront of this movement until the actual founding of the college in 1877 was the Reverend H. P. Jacobs. Rev. Jacobs was the founder and first president of the Mississippi Baptist Convention. Lelia Rhodes (1979, p. 14), who was commissioned to write the history of Jackson State, writes, "Jacobs was the spearhead of the idea and must be regarded, more than any other individual, as the founder of Jackson State."

DEVELOP PREACHERS, NOT SCHOLARS: EARLY INSTITUTIONAL LEADERSHIP

The early curricular and academic missions of most Black higher education institutions of the late nineteenth century reflect the views of Christian missionaries; in many cases, those of the American Baptist Home Missionary Society. Rhodes (1979, p. 18) comments, "The philosophy of the American Baptist Home Missionary Society was predicated on the conviction that the success of the Black population in the South depended upon capable self-government, moderated and guided by sound Christian ethics." True to the sentiments of most Northern philanthropic organizations, the American Baptist Home Missionary Society felt that Christian ethics were central to education not only for Blacks, but for Whites as well. Natchez Seminary intended to train Black ministers and schoolteachers. "The society had as its objectives the promotion and encouragement of moral life through the medium of religious education and the instruction and training of minority teachers who, in turn, would elevate their newly freed people to a useful and productive life" (Rhodes, 1979, p. 18).

The first president of Natchez Seminary was Dr. Charles Ayers, 1877–94. It was under Ayers's tenure that Natchez Seminary became Jackson College in 1883 and was moved to Jackson, Mississippi (Rhodes, 1979). In 1911, Dr. Zachary T. Hubert became the first Black president of Jackson College. He served in that capacity until 1927. Coombs (1973) describes the change in curriculum and scope of Jackson State beginning in 1913. President Hubert made a request to the Baptist Home Missionary Society to allow the institution to teach higher level courses. The request was denied. Coombs (1973, p. 35) states, "The American Baptist Home Mission Society was not ready for heresy. They wanted a school that taught Black people how to orate and preach—not one committed to the outlandish idea that Black people should study chemistry." Eight years later, in 1921, the society relented and allowed a more academic curriculum. In 1924, Mrs. Annie Mae Brown McGhee became the first graduate to earn a baccalaureate degree (Coombs, 1973). Transfer of control of Jackson College from the American Home Baptist Society to the Jackson College Board of Trustees occurred under the administration of Dr. B. Baldwin Danby in 1936. In 1940, under the leadership of then new president, Dr. Jacob L. Reddix, Jackson College began to receive legislated state support and became Jackson State College (Rhodes, 1979). In 1974, under the leadership of President John Peoples and the approval of Governor William Waller, Jackson State College became Jackson State University (Rhodes, 1979).

1970: REALITIES OF MISSISSIPPI RACE RELATIONS

In 1970, a main thoroughfare in Jackson, Mississippi, was Lynch Street. Ironically, it was named for John Roy Lynch, an ex-slave. Lynch was a self-educated man and served in Reconstruction era politics as a Mississippi state legislator and eventually a U.S. Congressman (Franklin, 1970). He also became an attorney, businessman, and Mississippi historian. He died in Chicago in 1939 at the age of 93 (Franklin, 1970). The irony lies in the fact that Mississippi had such a long history of racial hatred of Blacks, and it was from Lynch Street that Mississippi Highway patrolmen and Jackson city police fired into Alexander Hall and at a crowd of unarmed, fleeing students. Lynch Street was later renamed J. R. Lynch Street after the shootings in order to diffuse the tension associated with the tragedy and implied meaning of the word, lynch.

In order to keep the shootings in historical perspective, a review of several incidents that paved the way for May 14, 1970, is in order, as well as a brief look back at aspects of the racial and political climate of Jackson and the nation. In *Coming of Age in Mississippi*, Anne Moody relates her experience integrating a downtown Jackson Woolworth lunch counter in 1963. She describes the fear and violence of that day and how the Jackson police watched the sit-in participants get beaten and humiliated. Moody (1968, p. 267) writes, "About ninety policemen were standing outside the store; they had been watching the whole thing through the windows, but had not come in to stop the mob or do anything. . . . After the sit-in, all I could think of was how sick Mississippi whites were. They believed so much in the segregated Southern way of life, they would kill to preserve it."

In 1970, the Civil Rights Movement of the 1950s and 1960s remained fresh in the minds of most Americans, but many of the issues and ideologies were changing and becoming more radicalized (e.g., Black Power movement) (Joseph, 2006). In 1966, during the aftermath of James Meredith's shooting in Mississippi and the national attention it garnered, an emerging ideological division between the old and new civil rights leadership manifested itself with Stokley Carmichael's comments about Black Power. Carmichael and Martin Luther King Jr. clearly differed about strategies and approaches for social change, yet both men understood the power of organized political action, which was central to what Carmichael meant by using the militant sounding slogan. Historian Peniel Joseph (2006, p. 137) documents the characterization of the two by a *Newsweek* reporter in 1966 declaring King the "main-liner" and Carmichael the "hard-liner." Joseph (p. 146) comments, "King distanced himself from the slogan but refused to censure the meaning behind the message or the messenger." The struggle for equal treatment and opportunity for non-Whites continued in state and federal courts and at grassroots levels. The Vietnam War was still raging and in early May, 1970, President Richard Nixon announced the invasion of Cambodia. This action alone precipitated much of the Kent State student unrest and ensured that the war protests by many young Americans would increase. Several leading politicians ran campaigns and won elections on "law and order" platforms (e.g., Richard Daly in Chicago, Ronald Reagan in California, and Jim Rhodes in Ohio). That theme encapsulated the sentiments of a majority of American adults, especially White Americans, who seemed tired of the protests and complaints of those outside the mainstream (Goings, 1990).

Spofford (1988, p. 29) adds,

A week before the May '70 demonstrations began, California Governor Ronald Reagan, a candidate for reelection, had been asked about resolving campus unrest.

Reagan remarked: "if it takes a blood bath . . . let's get it over with." Ohio Governor James Rhodes, in a primary for the U.S. Senate, said the protesters at Kent State were "worse than the Brown Shirts and Communist element." In Washington, President Nixon called the demonstrators "bums" and Vice President Spiro Agnew called them "paranoids."

Clearly, protesting college students were viewed as "the enemy" and would be handled with militaristic tactics. It was not difficult to figure out what these men (and others in similar positions), who held the power to send armed troops to take actions in the name of restoring order, would do if and when the situation arose. In light of this national political climate, the racist legacy of the South adds another element of tension to what the historical context of the phrase "law and order" meant.

Jackson State existed in a climate of high racial tension and uneasy coexistence in the state capital (Rhodes, 1979). By most measures, Jackson State students in the late 1960s had not been excessively politically active. Of course, there were protests and incidents. However, Jackson State was not a hotbed of political protest when it came to student activism (O'Neil, Morris, and Mack, 1972). Weakley (1990, p. 65) states, "In 1970, the student body was not very involved in local or national politics. Although JSC had a few radical students, most students could only be considered modestly active at best."

Whether it was the brutal murder and mutilation of Emmett Till in the 1950s, the murder of Medgar Evers in 1963, or the little-known killing of Benjamin Brown by the highway patrol in 1967, Mississippi has a long and well-documented history of violence against Blacks. Brown was a young, Black Jackson resident killed by the highway patrol near the Jackson State campus under dubious circumstances (Spofford, 1988). These incidents, among others, were fresh in the collective consciousness of the students at Jackson State in 1970. The students were very aware of their surroundings and the history of their environment. The violent and racist reputation of the Mississippi Highway Patrol, Jackson city police, and a large segment of the community was a constant threat to the safety of any Black person who "stepped out of line" relevant to the Southern way of life.

THE SHOOTING GALLERY: LAW AND ORDER, MISSISSIPPI STYLE

Jackson State President Peoples characterized the throwing of rocks and bottles at White motorists passing through the campus on Lynch Street as the annual spring riot (Spofford, 1988). For a few consecutive springs in the years before 1970, groups of students and young, Black Jackson residents, known as "corner boys," gathered to throw rocks and bottles at passing cars. Why? President Peoples provided a synopsis of events surrounding the springtime riots and his first year at Jackson in 1967 (Spofford, 1988, p. 56):

> In May, barely two months after I assumed my duties, I had to deal with my first riot. The Jackson police chased a student driver onto the campus . . . They didn't catch the student, but they stopped at a men's dormitory where students were jeering them out of a window. One of the policemen fired a shotgun into the window, wounding a student football player in the face with bird shot.

May 13, 1970, seemed to signal another annual riot, except that there were new national and regional issues and frustrations on the minds of many Jackson State students. The Kent State shootings had just happened the week before, and the invasion of Cambodia was headline news all across the nation. Two days before, according to Spofford (1988, p. 34), "six blacks had been killed in a riot in Augusta, Georgia . . . All had been shot in the back . . ." Spofford (1988, p. 34) documented a callused quote from Georgia's Governor Lester Maddox, which illustrates the political climate of the times: "I've given orders to the troopers that if they're fired upon, that don't ask anybody to quit shootin', but blow whatever these people are in off their foundations if necessary to restore the peace." Spofford (1988, p. 36) continues, "With all these frustrations at Jackson State, small wonder that students revived the springtime ritual of throwing rocks at White drivers on Lynch Street. But they were not the only ones to toss rocks. To some, it looked as though a few corner boys had started it all." Henry Thompson, a student at Jackson State in 1970, comments,

> You have to vent your frustrations some way . . . These kids took them out on anyone that passed the college. It was like the whites were doing them an injustice for so long, and then they go and ride by in those nice big cars. It was like rubbing salt into the wounds. It was easy to throw rocks at them.
>
> (Spofford, 1988, p. 37)

To some students the Reserved Officers Training Corps (ROTC) building on campus seemed to represent all that the war and racist America stood for, and they did not support the covert implications. Spofford (1988, p. 35) adds, " . . . a rumor spread that Jackson State's ROTC barracks would go up in flames after dark. Many students wondered if this would be the night of the annual spring mini-riot."

Mississippi Governor, John Bell Williams, a former U.S. Congressman, was against civil rights and attacked President Truman and his integrationist policies on the House floor in 1948. Spofford (1988, p. 41) quotes one of Williams's race-baiting statements, "Chief among these are radical Negro, Communist and off-color mongrel organizations which, conceived in hate, whelped in treason and deceit, and nurtured on the breast of Communism, are attempting to bring about in this great country moral disintegration and mongrelization through a forced amalgamation of the races." Williams was a man cast in the mold of other blatantly racist Mississippi politicians, including James K. Vardaman (Governor: 1904–8) and Theodore G. Bilbo (Governor: 1916–20 and 1928–32; U.S. Senator: 1935–47).

On the evening of May 13, White motorists passing through the campus along Lynch Street were greeted with a barrage of rocks, bottles, and angry shouts. A police car was pelted with rocks and the officer called back to the station to make a report. Spofford (1988, p. 33) recounts the conversation between that police officer and the dispatcher from official transcripts this way: "Better tell them security guards out there they better get them niggers into them dormitories, or we fixin' to have some trouble out here . . . These niggers are congregated behind that fence, and . . . security guards should put them back in their rooms . . . They're throwin' . . . bottles and things over the fence into the street." Soon the police were on the scene in force and quickly blocked off Lynch Street.

Instead of the students calming down because of limited traffic through campus, some decided to try to burn down the ROTC building. The crowd continued to grow and became angrier. Campus security tried to get students to disperse, but failed. It soon became obvious to campus security that the ROTC building was a target of this growing rage. Meanwhile, a few cars had maneuvered past the barricade and were indiscriminately attacked with bottles and rocks. One of those cars was driven by Bert Case, a local newscaster, who had heard about the disturbance over the radio and came looking for the story (Spofford, 1988). Despite being warned by Police Chief W. D. Rayfield, Case drove through campus and was indeed attacked with rocks and bottles.

A call went to the governor's mansion from the police department. Spofford (1988, p. 41) states, "As prescribed by law, Governor Williams signed a proclamation authorizing the Mississippi Highway Patrol to join the city police in restoring order on campus." Governor Williams was an ardent supporter of law and order policies. When the call for the highway patrol was made, it brought to campus a group of men who had a long history of practicing violence upon the Black citizens of Mississippi. The stage was set for a disaster.

One police officer in particular showed up on the scene, Inspector Lloyd Jones, the unit's commander. Jones was known as "Goon" Jones by the Black citizens of Mississippi. He was the same man who in 1966 confronted Civil Rights marchers. Spofford (1988, p. 43) states, "When Martin Luther King Jr., Stokely Carmichael and thousands of other activists were marching from the Mississippi delta to the state capital to protest the recent shooting of James Meredith, Jones ordered the Civil Rights marchers tear gassed as they set up tents for sleeping." Meredith was the first Black American to integrate the University of Mississippi. "Goon" Jones also led highway patrolmen against a group of Black youth on Lynch Street in 1967, which ended in the death of Benjamin Brown. Jones later testified he had fired three times, but into the air (Spofford, 1988). Officer Jones was no friend to Black people.

To control the unruly students, law enforcement authorities in Jackson called in their heavy equipment: a military tank. The tank was known as "Thompson's Tank" and named after a former mayor who gave approval to the police department for its purchase. The tank's main purpose was riot control, although most Black citizens of Jackson knew it was only used against them. With a tank, over 20 policemen and a unit of the highway patrol, the campus was soon brought under control and the crowds dispersed. The ROTC building was threatened and some type of flammable object was thrown at it, but no damage was done and no one was injured (Spofford, 1988). At about 2:00 a.m., the students eventually returned to their rooms and the crisis was over. Or was it?

IN THE LINE OF FIRE

President Peoples invited student leaders to his residence to hear their concerns and learn reasons why the previous night's events occurred. The students mentioned the following reasons: the Vietnam War, the shootings at Kent State, the bell that was proposed to be installed on campus, and the presence of the ROTC building, which for many, represented the U.S. Military and all of the political uncertainties of the times (Young, 1990). Peoples listened and assured them he would look into their concerns, but made it clear that the actions of the night before could not be tolerated nor condoned. The student leaders left the president's residence, knowing that a mere conversation would not prevent further incidents by some of the angry students on campus or by the corner boys who visited the campus freely.

Peoples tried to reach an agreement with the mayor, Russell Davis, to barricade the street and divert traffic until things calmed down. Peoples (1990, p. 57) explains, "I pleaded with the city police authorities to keep up the street barricades so as to prevent any through traffic until we could be sure that the situation had cooled down . . . They said that there were a lot of people driving home from work who would be inconvenienced if they could not drive through Lynch Street."

Local newspapers might have contributed to the heightened confusion and intolerance of conflicts between Jackson State students and White residents of Jackson. *Jackson Clarion-Ledger* and the other local paper, *Jackson Daily News*, were owned by the Hederman family. Longtime Jackson residents, the Hedermans owned and controlled local newspapers, radio, and television stations (Spofford, 1988). The Hedermans supported the law and order sentiment of Governor Williams and promoted the Southern way of life. Spofford (1988, p. 54) states, "For decades, the Hederman press had shaped the unyielding segregationist ideology of white Mississippians." For example, after the 1963 March on Washington, *Clarion-Ledger* ran the following headline: "Washington is Clean Again with Negro Trash Removed" (Spofford, 1988, p. 55).

By the evening of May 14, students once again gathered around Alexander Hall, the women's residence building, socializing and sharing their feelings on the previous night's mini-riot. What began as a normal campus scene soon turned ugly. Around 9:30 p.m. a group of about 100 students gathered and cheered on rock throwers near Stewart Hall, a men's dormitory. Many of the rock throwers were corner boys who seemed to use the campus as a cover for their bad intentions, knowing that there would always be enough onlookers and perhaps helpers to get trouble going and keep it going (Young, 1990). The corner boys had nothing to lose and little to fear as part of the campus crowd.

Closer to Alexander Hall, a few students threw rocks at passing cars. Spofford (1988, p. 60) describes the event: "By 10:10 pm., a city policeman in a squad car spotted the crowd throwing rocks. Immediately he radioed headquarters: 'Call that security guard out there at Jackson State, and see if they can't scatter them niggers'." Around 11:00 p.m. some corner boys started a fire in a dump truck left overnight by workmen (Peoples, 1990; Spofford, 1988).

With a fire burning and the police alerted, the highway patrol and police were again summoned to help restore order. Lynch Street was barricaded by police and the fire department was called to extinguish the burning truck. The highway patrol and city police, escorting the firefighters, marched onto the campus and came face-to-face with hundreds of angry students. After the fire was extinguished, the fire trucks drove around the campus, avoiding the students, to handle another fire on the other side of the campus. In contrast and in a show of force, law enforcement officers decided to march straight through the campus. They stopped in front of Alexander Hall where a large group of students had gathered (Peoples, 1990). A major problem at the scene was the lack of coordination and communication between the police and the patrol; no one was clearly in charge, nor was there a clear plan of action (Spofford, 1988). The officers were as much a loose mob as the students, except that the officers were heavily armed and willing to use deadly force.

With students and law enforcement officers facing off in front of Alexander Hall, a standoff was imminent. Thompson's tank was maneuvered in front of the dormitory and both police and patrol units took positions in front of the jeering students. Lt. Magee, a city policeman, used a bullhorn to order students to disperse. Shouts of "pig" and "whitey" were heard as the stage was set for the tragic moment (Spofford, 1988).

The women residents of Alexander Hall were busy attending to the typical activities of college students living in a residence hall: reading, studying, listening to music, and visiting one another. Many residents were unaware of the events taking place outside or of the danger lurking across the street. Some students, aware of the ruckus in front of the building, watched from their windows. Alexander Hall has a five-story tower stairwell facing Lynch Street. The police lines were directly in front of this part of the building.

Students did not disperse as Magee ordered and a few rocks were thrown at the officers. Suddenly a bottle was thrown from the crowd of students. It landed behind the officers with a sharp crashing sound. As soon as the sound of the crashing bottle filled the air, the officers began firing at will with shotguns and automatic rifles. Weakley (1990, p. 66) states, "The moment the bottle hit the ground the police and highway patrolmen appeared to go crazy. They began to fire their weapons as if all they had been waiting for was an excuse to fire." The students, caught completely off guard, began running frantically. Gunfire flashed through the night. Students dropped to the ground like soldiers in a movie. Many students scrambled wildly toward the doorway of Alexander Hall and began piling up in the doorway, trying to get out of the line of fire (Weakley, 1990). The gunmen continued to fire their weapons into the crowd and into the building. The building was shot full of holes. Residents were wounded by bullets, buckshot, and flying glass while sitting in their rooms, running into the halls, or crouching in the lobby. The gunfire lasted approximately 30 seconds (Spofford, 1988; Weakley, 1990). Both police and patrolmen reported that there had been a sniper inside the dormitory, although no proof of a sniper was ever found (Spofford, 1988; O'Neil et al., 1972). This "sniper" story was strangely similar to the one given at Kent State to justify that shooting.

After the ambulances took the injured away and the law enforcement forces left the campus, students refused to go to their dorm rooms (Spofford, 1988). President Peoples arrived on the campus and tried to calm the crowd and ascertain what had happened (Young, 1990). Students gathered blankets and sat out on the lawn of Alexander Hall the rest of the night, crying, praying, and asking themselves what had happened (Young, 1990).

TWO SURVIVORS' STORIES

After the smoke cleared, two were dead and at least 12 wounded. These survivors' stories are understood within the framework of CRT (Saddler, 2004; Ladson-Billings,

2005). The stories, although brief, add to the robust context of the onslaught and its aftermath (Ladson-Billings, 2005). Their words illuminate the overt and underlying racial oppression of the event and their status as potential double-victims who understood how they might easily become further victimized even after the smoke had cleared. Weakley was one of the students wounded that night. He writes,

> I was in a state of shock, cold and trembling violently as I lay on the ground. Though I knew I had been wounded, it didn't hurt, didn't even seem to matter. I could feel my pants leg wet with blood. Then the cold feeling was replaced with a warm tingling sensation in my leg. I saw Howard Levite, one of my fraternity brothers, peering out from a door inside the dorm. I screamed out to him to help me. How he distinguished my voice from the others, I don't know. I could tell from the look on his face that he was also in a state of shock, looking out over the mass of bodies on the ground. Though I'm sure he was afraid, he was the first person to stand up after the shooting and he moved toward me, stepping over people who lay in the doorway. To this day–though until now I have never been able to share this with him, or with anyone–I admire him for the courage he showed in the face of danger. I was terrified that the highway patrolmen, still a few feet behind me, would kill Levite. I wanted to scream to him to go back, but the words would not come out of my mouth.
>
> (Weakley, 1990, p. 67)

Even after shooting unarmed, fleeing college students, highway patrolmen spewed race hate and bad intentions. Weakley (1990, p. 67) relates what happened after his friend reached him:

> A big burly highway patrolman pointed his weapon at us and said "Nigger, you'd better stay your ass on the ground" . . . Another patrolman came over and said, "leave those niggers alone." Pointing, he continued, "There's a dead nigger over there, a more seriously wounded one next to him."

Gloria Mayhorn, a student wounded by gunfire inside the dorm, shares an account of what happened to her:

> "They're shooting rice," I was thinking, but there were pellets hitting my body. They were stinging they just had me covered . . . Then I felt a big prick it felt like a bee sting and I felt blood running down my arm. I looked and there was a perfect hole on one side, and the other side was blasted out. "I'd better get out of here," I thought to myself. I started scrimmaging my way out of there and I slipped I guess on blood. I lost my beach thongs. I was on my hands and knees at the first step in the stairwell, and then when I raised up to continue, I felt what I thought was a shot hit my head, and another singed my back. Something hit me on the right side of the back of the head. Blood was pouring like from a faucet, and I thought if I ran, it would flow faster. I looked as if I had measles from all the pellet marks. I had glass cuts in my face and shoulders. Glass was in my hair.
>
> (Spofford, 1988, p. 72)

Over 250 shots were fired (Spofford, 1988). It was a miracle more people were not killed or injured. Alexander Hall looked like a shooting gallery. Bullet holes as large as silver dollars remain in the concrete walls. Most of the windows of the stairwell and the entire side of the building were shattered. A power line outside of the building was cut in two by the firepower (Spofford, 1988). Just as suddenly as it began, it was over. Thirty seconds of blind hate and rage left a permanent scar on those injured, the families of the two men killed, on Jackson State, and the nation.

Ambulances were called and the wounded were taken to the hospital. The nearest hospital refused to treat the students because they were Black; so they had to be taken across town to another facility (Spofford, 1988). Regardless of the horror and magnitude of the incident that just occurred, this was still Mississippi and overt racism was a constant reality. Several students reported being insulted and ignored for over an hour by nurses and doctors in the emergency room (Spofford, 1988). Many of those wounded were hit with buckshot. Some refused to be taken to the hospital out of fear that they would be injured further. Weakley (1990) reports that he went to his family doctor instead of the hospital, as did many others. He contends that close to 40 students were actually wounded or injured, but did not get identified for the official count (Spofford, 1988; Weakley, 1990). Only those treated at the hospital that night were counted as wounded.

One of the dead was Phillip Gibbs, a junior at Jackson State. The other was James Earl Green, a high school senior who was walking through campus on his way home after work (Young, 1990; *New York Times*, May 15, 1970). Gibbs, a husband and father, was visiting friends on campus when the shooting started. He was shot in front of Alexander Hall.

Green lived close to the campus and always used the campus grounds as a shortcut to get home after working at a local grocery store. He was shot in front of Roberts Dining Hall, opposite to Alexander Hall (Huge, 1970). Police turned and fired at Roberts and Stewart Halls, which were down from Alexander Hall, after the initial attack on Alexander Hall (Huge, 1970). All three buildings were marked with holes from bullets and shotgun blasts. Green's mother was worried that he was late getting home from work and learned of her son's tragic death while watching television.

THE AFTERMATH, MEDIA ATTENTION, AND INVESTIGATIONS

Huge (1970, p. 67) describes the damage to the buildings in graphic detail in the following passage:

> In some places there were holes the size of grapefruits. . . . The glass was shattered on every floor, and the blue-green, double thick steel panels at the base of every landing were also perforated. There were two types of bullet holes in those steel panels one, which had been caused by rifle fire, perhaps the high velocity fire of automatic weapons. Those bullets had come through both sheets of steel so that one could look through the two holes onto the street where the rifleman stood as he fired up at the girl's dormitory. The other most frequently used weapons were shotguns with "Double OO" pellets, which are supposed to be able to kill a deer at 100 yards.

Some limited national attention was given to the incident (*New York Times*, May 15 and 16, 1970). Many colleges and universities around the country closed for the rest of the spring, but mostly due to the tragedy at Kent State. President Richard Nixon sent Attorney General John Mitchell to the Jackson State campus to meet with Peoples and requested the Federal Bureau of Investigation (FBI) to investigate (O'Neil et al., 1972; *New York Times*, 1970; Spofford, 1988). A task force of U.S. Congressmen held hearings to gather information (O'Neil et al., 1972). The state of Mississippi and federal government (President's commission on Campus Unrest-Scranton Commission) conducted hearings to determine the facts of what occurred (O'Neil et al., 1972; Spofford, 1988).

Two years after the shootings, a civil trial was held in Biloxi, Mississippi. The families of Gibbs and Green as well as those wounded sought over $13 million from the law enforcement agencies involved in the shooting. An all-White jury found for the defendants (Spofford, 1988). One young lady, wounded at Jackson State, attended the trial in Biloxi and stated that killing Blacks did not matter in Mississippi; nothing will ever come of it anyway (Spofford, 1988). Did law and order truly prevail, or was the incident another chapter in the racial narrative of America? Clearly race played a significant factor, and the notion of social justice was ill served prior to and during the incident. In the final analysis, no one was ever held accountable or responsible for the shooting and killing of Black citizens. Eventually, a pleasant-looking plaza was built on the campus, which cut off Lynch Street to traffic. The plaza was named Gibbs–Green Plaza as a memorial to the two young men killed on that tragic night. Examining this tragic incident through the lens of CRT adds a race-conscious perspective that has increasingly become lost when "the ideal of color-blindness became the official norms of racial enlightenment" (Crenshaw et al., 1995). If race and racism were never central to the African American experience and the experiences of other people of color in the United States, then using a CRT lens implodes under its own philosophical weight. However, if we acknowledge that race has been and remains a critical factor in the social, cultural, political, legal, and economic fabric of America, then using CRT as one of many possible lenses through which to analyze the educational and historical experiences of people of color is not only necessary, but essential. The violence perpetrated against the students at Jackson State had its roots in racism and in the high-stakes political climate of the late 1960s. The direction of how higher education would be managed and controlled faced distinct challenges by 1960s student protest and activism, and the power structures that monitored and enforced order over those institutions made loud and clear statements about what would and would not be tolerated. Race became an additional complex factor in the systemic response to student protest at Jackson State on May 14, 1970.

CONCLUSION

The tragedy at Jackson State is one example of how certain histories of higher education are not shared as widely as others. Given the racial history of the United States, it matters greatly as to the race and social status of victims of violence or tragic actions as to the attention or memory allotted to a particular incident or situation. For many Blacks, and other racial or ethnic minorities, gaining an education, especially at the collegiate level, is a struggle fought on a cultural battlefield of ideological, economic, and social landmines that they feel they have little power to influence. Learning about, understanding, and analyzing that historical struggle should contain a consideration of race and racism because it has long been a stark, daily reality for those deemed as the "other." CRT offers "a discourse of liberation, [that] can be used as a methodological and epistemological tool to expose the ways race and racism affect the education and lives of racial minorities in the United States" (Parker and Lynn, 2002, p. 7). How can we view historical events through race-blind perspectives when race-conscious customs, laws, and behaviors shaped the very fabric of America and American education (Brown et al., 2003)? CRT can help advance social justice agendas by addressing "the ways race and racism are deeply embedded within the framework of American society" (Parker and Lynn, 2002, p. 8).

The shootings at Jackson State, viewed through a CRT lens, lays bare the historical fractures and injustices of racism and White privilege that has shaped and continues to influence many of the popular narratives of higher education, who matters most, and why. Principles of social justice must prevail as we review and understand the context in which HBCUs were created, changed over time, and operate in the twenty-first century.

Many people have heard of the Kent State shootings, even if they are not old enough to remember it. Most know nothing about the killing of Black students on Black college campuses between 1968 and 1970, and, in particular, the shooting of Black students ten days after Kent State. The teaching and researching of higher education history is ill served if only stories of one particular group or of one type are told and honored in exclusion of others.

REFERENCES

Brown, M. K., M. Carnoy, E. Currie, T. Duster, D. B. Oppenheimer, M. Shultz, and D. Wellman. 2003. *White-washing race: The myth of a color-blind society.* Berkeley, CA: University of California Press.

Coombs, O. 1973. The necessity of excellence: Jackson State college. *Change* 5(8): 34–9.

Crenshaw, K., N. Gotanda, G. Peller, and K. Thomas, eds. 1995. *Critical race theory: Key writings that formed the movement.* New York: New Press.

Delgado, R., ed. 1995. *Critical race theory: The cutting edge.* Philadelphia, PA: Temple University Press.

Dionne, E. J. 1990. Jackson state remembers. *Washington Post*, May 4, A1.

Franklin, J. H., ed. 1970. *Reminiscences of an active life: The autobiography of John Roy Lynch.* Chicago: University of Chicago Press.

Goings, K. W. 1990. The Kent State tragedy: Why did it happen? Why it could happen again? *Education Digest* 56(2): 57–60.

Harrison, E. C. 1972. Student unrest on the Black college campus. *Journal of Negro Education* 41(2): 113–20.

Huge, H. 1970. Inquest at Jackson State. *New South* (Summer): 65–70.

Joseph, P. E. 2006. *Waiting 'til the midnight hour: A narrative history of Black power in America.* New York: Henry Holt.

Kennedy, E. M. 1970. Student riots, civil rights, and the rule of law. *Social Science* 45(2): 6–9.

Ladson-Billings, G. 2005. The evolving role of critical race theory in educational scholarship. *Race Ethnicity and Education* 8(1): 115–9.

Ladson-Billings, G., and W. F. Tate. Fall, 1995. Toward a critical race theory of education. *Teachers College Record* 97(1): 47–68.

Lesher, S. 1971. Jackson State a year after. *New York Times,* March 21, SM24.

Lucas, C. J. 1994. *American higher education: A history.* New York: St. Martin's Press.

Moody, A. 1968. *Coming of age in Mississippi.* New York: Dell.

Nixon voices regret in the deaths of 2 Negro youths in Jackson. 1970. *New York Times,* May 17, 61. (wire service)

O'Neil, R. M., John P. Morris, and Raymond Mack. 1972. *No heroes, no villains: New perspectives on Kent State and Jackson State.* London: Jossey-Bass.

Parker, L., and M. Lynn. 2002. What's race got to do with it? Critical race theory's conflicts with and connections to qualitative research methodology and epistemology. *Qualitative Inquiry* 8(1): 7–22.

Peoples, J. A. 1990. The killings at Jackson State University May 1970: Reminiscences of Dr. John A. Peoples, President, Jackson State University, 1967–1984. *Vietnam Generation* 2(2): 55–8.

Reed, R. 1970. F.B.I. investigating killing of 2 Negroes in Jackson. *New York Times,* May 16, 1–15.

Rhodes, L. G. 1979. *Jackson State University: The first hundred years, 1877–1977.* Jackson, MI: University of Mississippi Press.

Rosenthal, J. 1975. Southern Black student activism: Assimilation vs. nationalism. *Journal of Negro Education* 44(2): 113–29.

Saddler, C. 2004. The impact of Brown on African American students: A Critical Race Theoretical perspective. *Educational Studies* 37(1): 41–55.

Spofford, T. 1988. *Lynch street.* Kent, OH: Kent State University Press.

Thelin, J. R. 2004. *A history of American higher education.* Baltimore, MD: Johns Hopkins University Press.

Weakley, V. S. 1990. Mississippi killing zone: An eyewitness account of the events surrounding the murders by Mississippi highway patrol at Jackson State College. *Vietnam Generation* 2(2): 113–20.

Williams, J., and D. Ashley. 2004. *I'll find a way or make one: A Tribute to historically Black colleges and universities.* New York: Amistad.

Winant, H. 2002. Theoretical status of the concept of race. In *Theories of race and racism: A reader,* eds. L. Back and J. Solomos, 181–90. New York: Routledge.

Wise, T. J. 2005. *Affirmative action: Racial preference in Black and White.* New York: Routledge.

Young, G. C. 1990. May 15, 1970: The miracle at Jackson State College. *Vietnam Generation* 2(2): 75–81.

CHAPTER 8

ON OPPOSITE SIDES
OF THE TRACK:
NEW ORLEANS' URBAN
UNIVERSITIES IN BLACK
AND WHITE

VALERA T. FRANCIS AND AMY E. WELLS

The history of the urban comprehensive university in metropolitan New Orleans stands out as an example of segregated institution-building in the years after *Brown v. Board of Education* (*Brown v. Board of Education of Topeka*, 1954). This initial history involves the establishment of two universities just two miles apart in the New Orleans Lakefront area: Southern University at New Orleans (now SUNO), chartered in 1956 and opened in 1959 as an "extension" of Southern University in Baton Rouge (SU) and part of what is now the nation's only historically Black university state system, and the University of New Orleans (UNO), a part of the Louisiana State University (LSU) system, formerly named LSUNO and founded in 1958.

Although the UNO touts itself today as the "first racially integrated, public university in the South" (University of New Orleans, n.d.) and historians credited Governor Earl K. Long for the "peaceful opening of a fully desegregated university" (Kurtz and Peoples, 1990), this research uncovers a more complicated history involving campus protest and political maneuvering to spoil enrollment for "unwanted matriculants [*sic*]," ("LSU Board 'Taunts' 53 Negro Students," 1958). Our story demonstrates that SUNO arose from the state's intent to flirt with but not fold to federal and local demands for integration in education. To this end, Whites considered SUNO far more important than Blacks for preservation of segregation and forced *their* will through politics and the public purse.

SETTING: NEW ORLEANS, LOUISIANA

In 1954 New Orleans, a troublesome predicament included the fact that there were no state-supported public educational institutions at the college level in the city, for Black or White students. With the land-grant campuses of LSU and Southern University (for Negroes, established in 1890) 90 miles away and Southeastern University 55 miles north in Hammond, the State Board of Education (SBOE) began to study the need and feasibility of remedying the situation by establishing two separate and segregated commuter colleges in New Orleans (*Resolution*, 1954).

To underscore the importance of this issue, the Young Men's Business Club of New Orleans, an organization of prominent White citizens, endorsed the idea on March 31, 1954 (*Resolution*, 1954). As early as May 20, 1954, two days after the *Brown* decision, state representatives introduced three bills to create two separate public institutions of higher education. The first of the two bills, No. 562, introduced by four representatives, called for

> a school for higher education in the arts and sciences for white children of the State of Louisiana, in the New Orleans area and under supervision of the State Board of Education; to provide for the building, equipping and maintenance of said institution; and repealing of all laws in conflict herewith.

Perhaps, signaling its importance, 16 legislators then introduced House Bill No. 563, which sought

> to create and establish a school for higher education in the arts and sciences for colored children of the state of Louisiana, in the New Orleans area, under the supervision of the State Board of Education; to provide for the building, equipping and maintenance of said institution; and repealing all laws in conflict herewith.

Representative James Beeson of Jefferson Parish who introduced bills 562, 563, and 564, argued for approval: "We believe that if there is any section of the state deserving a college, it is this area. We have 65 percent of the population, pay more than that in severance and sales taxes. We don't want a dormitory college and don't want to take in outside students." Representative Beeson estimated that the total cost of the two schools would be $3,800,000 ("Colleges Given House Unit Okay," 1954).

Representative Beeson's original and modest plan for institutional governance and finance of the proposed commuter colleges in New Orleans grew into a political football carried into office by Governor Long. Long, brother of former governor Huey Long and heir to the Long family's populist tradition, had served two previous terms as governor (1939–40; 1948–52) and promised on the campaign trail for his third term (1956–60) an affordable option for metropolitan area residents desiring to attend school in Baton Rouge but unable to commute or afford room and board charges (Kurtz, 2001; Kurtz and Peoples, 1990).

In 1956, when Louisiana voters sent Long back into the governor's mansion, half of the 208 formerly all-White colleges and universities in the 17 Southern and border states had opened their doors to Negroes at some level. Of the 17, only four states, Florida, Georgia, Mississippi, and South Carolina, had yet to admit Blacks to their publicly state-supported, all-White institutions (Wallenstein, 1999). In Louisiana, starting with the admission of Negro graduate students at LSU in 1951, three more of the state's seven all-White public colleges had enrolled Blacks at the undergraduate level following the *Brown* decision ("Tax Backed Colleges Open Doors," 1956). There were 34 state-supported institutions for Negroes in the 17 Southern and border states with a combined enrollment of 48,168 students (Clark, 1958).

Neither the *Brown* ruling nor the change in administration did much to slacken the state's official position on keeping public schools segregated. As a successful populist, Governor-elect Long had courted and won the Black vote while simultaneously pledging that he would do everything in his power to close any public school where integration was ordered either by federal court or by an act of Congress (Kurtz and Peoples, 1990; Fenton, 1957). In addition to Long and his pandering, Senator William Rainach won reelection to the state senate without opposition and continued his work as the chair of the Joint Legislative Committee on Segregation (JLCS) ("Louisiana Governor Pledged to Ban Schools If Courts Act," 1956).

Under the mantle of state's rights, White southerners and segregationists maneuvered to stall integration and impending social change (Bartley, 1969; Orfield, 1969; Vander Zanden, 1958; 1959; 1962). Despite the *Brown* decision, barriers to desegregation in the publicly tax supported institutions of higher education in Louisiana remained essentially legal. Reversal of federal judicial action that had opened doors of White institutions to Negroes in the state (*Wilson v. Board of Supervisors,* 1950) as well as prevention of desegregation at the undergraduate level topped the segregationists' education package. Another tactic involved continuing to build segregated institutions in a *Plessy*-inspired defiance of the Supreme Court (*Plessy v. Ferguson,* 163 U.S. 537, 1896) arguing, of course, that Blacks truly desired to attend their *own* blatantly unequal institutions.

Act 15 stood out as one centerpiece of segregationists' efforts. Proposed by the JLCS, *Act 15* required that students entering institutions of higher education supported by the state be certified with good moral character by principals and superintendents before they could be enrolled and furthermore, school officials who aided in desegregation of state educational institutions would be subject to removal ("100 Negroes Registered in Louisiana Colleges Under Courts' Injunction," 1957). This basic measure prevented additional Blacks from registration and enrollment in four formerly all-White public institutions of higher education, namely, LSU, Southwestern Louisiana Institute, Southeastern Louisiana College, and McNeese State College; it also forced out of these institutions those Negroes already enrolled as a result of judicial order. With the adoption of *Act 15*, Louisiana became the first state to attempt to resegregate its colleges and universities after the *Brown* decision ("College Segregation-Desegregation Issue Revived by New Court Action," 1957).

Within this context the new Negro commuter college took shape as a branch of Southern University in the 27th Extraordinary Session of the Legislature for Louisiana that convened on August 30, 1956. On the first day of the session, Messrs. Cashio, Holt, Bertrand, and Bosetta introduced House Bill No. 27, "An Act to establish, as a branch, or extension of Southern University a Negro College in the Greater New Orleans Metropolitan area, to provide for its administration; and to appropriate One Million Fifty Thousand Dollars ($1,050,000) out of the General Fund of the

State of Louisiana for the fiscal years 1956–1957 and 1957–1958 for its establishment and maintenance." On August 31, it was read a second time and referred to the Committee on Appropriation, which reported favorably on the act. The following day, the act was returned to and called from the Calendar of the Legislature and read a third time before the full legislature. On the roll call for final passage, there were 69 yeas and 9 nays. Having received a two-thirds vote of the members elect, the bill passed and moved on to the Senate. On September 4, by a vote of 62 yeas and zero nays, House Bill No. 27 gained the Senate's unanimous approval in open session and expedition to the governor for executive approval (State of Louisiana, *Official Journal*, 1956; State of Louisiana, *Acts of the Legislature*, 1956).

Although the LSU lobby stalled the idea of any LSU branch campus in New Orleans for some time, Long eventually prevailed in one of the bloodier battles of his career. For the site of what had become an LSU branch campus in New Orleans, Long had his eye on the soon-to-be abandoned Camp Leroy Johnson in the Pontchartrain Lakefront area, land controlled by the very powerful New Orleans Levee Board. Backed by the New Orleans press, Chamber of Commerce, and city hall, the Levee Board persisted in plans to level the air station buildings and subdivide the 200-acre facility for residences. Long's plan appeared doomed, except that he used his gubernatorial appointment power to dismiss the members of the Levee Board and reappoint new patrons keen on his plan to approve a college moving in as soon as the navy moved out. The new LSU New Orleans (LSUNO, later UNO), approved by the Louisiana Legislature in June 1958 with an appropriation of $8.1 million, opened that September (Kurtz and Peoples, 1990; "Uneventful Louisiana College Integration Involves 160–175," 1957).

THE PLANNING FOR SUNO BEGINS

The planning and opening of SUNO evolved slowly and on contested ground. As stipulated by *Act 28*, "the State Board of Education was vested with full authority to determine the location of and acquire the land and buildings determined necessary for the establishment of the college." Operating as provided by the Constitution of 1921, this board administered the affairs of all public institutions of higher education in the state except for the LSU System. Eleven members, without required professional or academic qualifications, male or female, and from any and all walks of life, made up the board's membership. In the 1950s, all 11 members were elected by popular vote. They served terms based on districts; one each from the congressional districts for terms of eight years, and one from each of the public service districts for terms of six years (Committee of the Faculty of the College of Education, 1969).

The board administered the affairs of eight state colleges and universities, including Grambling College (now historically Black), Louisiana Polytechnic Institute, McNeese State College, Francis T. Nichols State College, SU, special schools, and trade schools. In this capacity, the board authorized purchases of land, buildings, and equipment as well as provided for all repairs, constructions, and improvements needed by the colleges, special schools, and trade schools. For these purposes, the board incurred debts and issued notes, bonds, or certificates of indebtedness. The eleven-member board served as the governing board for the new Negro institution, and assumed responsibility for its direct operation (Committee of the Faculty of the College of Education, 1969). As was customary at the time, all members of the board were White males.

With such extensive jurisdiction and control, the board proceeded with plans to establish a Negro institution regardless of federal efforts to integrate. The board emphasized the positives for Negroes, and the city's direct benefit and the location of this new facility remained crucial. The board immediately appointed a special committee to begin research, and for the remainder of the year the committee concentrated on finding an appropriate site. By the committee's January 25, 1957, meeting, Nash Roberts, its chairman, touted the suitability of the Pontchartrain Park Homes Subdivision in the Gentilly area (State Department of Education, 1957), the first subdivision in the city designed for African Americans when other developments specifically excluded Blacks. In order to quell the restiveness of a growing Black middle class, New Orleans' Mayor Chep Morrison led the charge to set aside 200 acres of land in the new "super-deluxe" subdivision for African Americans ("Dream House," 2003).

The state offered two tracts of approximately 20 to 30 acres each, at the northwest and southwest corners of Pontchartrain Park located along Press Drive near an 18-hole golf course. Either would be well suited for such a facility in this subdivision that had won national recognition as an outstanding Negro development ("Option Steps for Pontchartrain Site for College Branch Okayed," 1957). After all, argued Roberts, the use of land for such a needful purpose would be mutually beneficial to the city and Black citizens. Soon the board empowered Nash's committee to negotiate any options necessary to obtain this site for the Southern University's New Orleans branch (State Department of Education, 1957).

Four months later, Roberts reported the committee's progress. At this meeting, the board agreed to purchase 5.407 acres of land from the city and 1.141 acres of land from Pontchartrain Park Homes, Inc. at the appraised value of the New Orleans Real Estate Board (State Department of Education, *Bulletin No. 842*, May 3, 1957). In July 1957, Louisiana's superintendent of education authorized the preparation of the act of sale for the 6.8-acre portion of Pontchartrain Park, which the state purchased for the city as part of the campus of the proposed four-year Negro college. The city sold the land for $700 per acre with the understanding that most of the proceeds would go back into providing utilities extensions to the institution. The college would be able to use the baseball and football stadium in the park for its own sports events and the park itself could be utilized as a part of the campus area ("State Prepared to Acquire Site," 1957). Thus, the board planned to use the $50,000 appropriated in *Act 28* to acquire the property (State of Louisiana, *Bulletin No. 846*, 1957). With the legislature's appropriation of $1,050,000 to get this Negro college started, nothing stopped the proposed project that endorsed segregation rather than integration of Louisiana higher education.

Meanwhile, as Louisiana forged ahead with the plan of building a new institution for Negroes in 1957, desegregation plans had been adopted for education at all state-supported institutions in Oklahoma, Missouri, Arkansas, Kentucky, West Virginia, Maryland, and Delaware, with partial desegregation occurring in some public colleges and universities in Texas, Tennessee, Virginia, North Carolina, and even in Louisiana (Kurtz and Peoples, 1990; "College Segregation-Desegregation Issue Revived by New

Court Action," 1957). Segregation in higher education remained intact in Alabama, Mississippi, Georgia, South Carolina, and Florida (Johnson, Cobb-Roberts, and Shircliffe, 2007; Wallenstein, 1999). In these states, Negroes in relatively small numbers had entered some institutions while none had matriculated at others, with some White students entering schools formerly maintained for Negroes (Wallenstein, 1999).

THE STAKES WERE HIGH FOR LSU

The LSU Board of Supervisors closely monitored developments related to the progress of the Southern branch in New Orleans. The board had long fought to keep the LSU system segregated, dating back to the interstate compact agreements created in the mid-1940s by the Southern Regional Education Board (SREB) when the state participated in the compact to keep Blacks from attending its graduate and professional schools at LSU (Haskew, 1968; Sugg and Jones, 1951; Thompson, 1949; Wells, 2002; 2004). Only legal action forced the university to admit Negroes to its law and graduate schools in the early 1950s (*Payne v. Board of Supervisors*, 1950; *Wilson v. LSU Board of Supervisors of Louisiana State University*, 1950).

By 1953, LSU's segregated undergraduate programs came under attack with the enrollment of A. P. Tureaud Jr., son of New Orleans' lead attorney for the National Association for the Advancement of Colored People (NAACP). Tureaud attended LSU for a short time, living in a university dormitory, before LSU revoked his registration on the grounds that equal facilities at SU existed ("Louisiana," 1954). Following a series of court rulings in the case, the U.S. Fifth Circuit Court of Appeals ordered LSU to admit Tureaud. Following the court's ruling, LSU President Troy Middleton stated that young Tureaud would be admitted "if he makes application and is found to be qualified." Middleton added, "The doors will not be opened, however, to all Negroes who desire to enter the undergraduate schools, and the university still has the right to apply to the Supreme Court for a writ of review" (*Southern School News*, "School Issue Sharpened in Louisiana as New Year Begins," September 1955). The tall, Mississippi-born and -educated Middleton, a World War II combat general, personally favored racial separation in public schools and worked to preserve it ("Legislature Investigates Pro-Integration Feeling in Nine Colleges," 1958).

In the years between the court's decision in 1955 and 1958, the LSU Board of Supervisors remained committed to segregation at the undergraduate level. The idea of a commuter college, an LSU branch in New Orleans, had been discussed in the legislature for years ("Uneventful Louisiana College Integration Involves 160–175," 1957). While the board kept the undergraduate programs on the Baton Rouge campus free of Negro students, it remained concerned over whether its new commuter college in New Orleans, scheduled to open in September 1958, would be forced to register and admit Negroes to the freshman class.

As the public debated the problems of this new institution, such as the firing of the board opposing site use or the wisdom of starting another state college so close to two existing ones, few mentioned segregation. Although LSU had been under an integration order at the undergraduate level since the January 1956 court ruling in *Tureaud v. LSU Board of Supervisors*, whether the new LSU branch would be covered by the ruling lingered as an debatable question to the board ("Uneventful Louisiana College Integration Involves 160–175," 1957). NAACP attorney A. P. Tureaud Sr., father of the principal in the litigation, stated, "This order applies to all branches without exception. The day the LSU branch opens in New Orleans, Negroes will be there to register." He went on to say "this would hold true even if the Negro college branch proposed for New Orleans at the 1956 legislature is built" ("Uneventful Louisiana College Integration Involves 160–175," 1957).

By April 1958, the LSU Board of Supervisors decided not to admit Negroes to the new branch at New Orleans without a court fight. When the registrar's office opened on April 1, approximately 75 Negroes applied for admission and all were turned down. Registrar W. R. Beeson told each Negro applicant that he acted in accordance with state law and the policies of the LSU Board. On July 29, 11 of the students filed suit in federal district court (*Henley et al. v. Board of Supervisors of LSU*, 1958) and claimed they were deprived of their civil rights. U.S. District Judge Herbert W. Christenberry of New Orleans, urged by Negro counsel for a speedy decision, granted a preliminary injunction against LSUNO's announced policy on September 8, 1958— the first day of registration at the new LSU branch. As a result, 59 Negroes comprised the enrollment of LSUNO's first freshman class of 1,500 students.

On behalf of the state, Attorney General Jack Gremillion appealed Judge Christenberry's ruling in the U.S. Fifth Circuit Court. The state asked for a stay of the decision of the lower court, but Appellate Judge John Minor Wisdom, speaking for the three-member court, said, "No showing having been made to justify granting the application for the stay, the motion is denied" ("New LSU Branch Opens with Classes Mixed under Federal Court Edict," 1958). The appeals court decision came four days after Christenberry's ruling; and the 59 Negroes who had registered began attending classes in the old Naval Air Station buildings of LSUNO.

Although considerable racial tension permeated the opening of LSUNO, no incidents marked the first week of classes ("New LSU Branch Opens with Classes Mixed under Federal Court Edict," 1958). The LSU Board of Supervisors added to the tense atmosphere when, on September 13, they met and declared Negroes "unwanted" and warned that their continued attendance at LSU or its branches resulted in "enmity, instead of the feeling of mutual respect" ("LSU Board 'Taunts' 53 Negro Students," 1958; "New LSU Branch Opens with Classes Mixed under Federal Court Edict," 1958). The board further declared, "This board wishes to point out that any Negro student whose enrollment is forced upon this university, enters as an unwanted matriculant [*sic*]" ("LSU Board 'Taunts' 53 Negro Students," 1958). This statement stood out as the strongest salvo launched by the LSU Board in the segregation fight.

Minor incidents soon interrupted the calmness that prevailed during the first week of classes. Just two weeks after the first Negro students gained admission to LSUNO, one Black male was expelled for carrying a concealed sheath knife; police booked a 46-year-old Negro minister for loitering; Ku Klux Klan (KKK) flags were flown from the 80-foot campus smokestack, and anti-Negro signs were painted on two buildings. Against the background of a sign posted warning Negroes of their unwanted status, a cross burning occurred just outside the campus grounds ("New LSU Branch Opens with Classes Mixed under Federal Court Edict," 1958; "LSU 'Taunts' 53 Negro Students," 1958).

Dr. Homer Hitt, newly appointed dean of LSUNO, responded quickly. He issued a warning to students, which said in part: "The unfortunate incidents of the last two days (September 16–17) make it necessary for me to remind you that LSUNO expects and will demand orderly behavior of its students. We do not intend to permit our academic program to be disrupted by boisterousness and violence" ("New LSU Branch Opens with Classes Mixed under Federal Court Edict," 1958). Hitt made it clear that any students involved in such behavior risked suspension or expulsion from the institution ("One Student is Expelled in Incidents," 1958).

The opening of the integrated LSUNO did not go without complaint in New Orleans' White community. For example, the Gentilly Citizens Council wired the LSU Board asking that the university be closed, stating, "Whereas, the LSUNO is located in our immediate vicinity and realizing it will only be a short time until the school will be all Negro, we urgently request you to close the school forthwith. The people of Louisiana have made it clear that they do not want integration in any shape or form" ("New LSU Branch Opens with Classes Mixed under Federal Court Edict," 1958). Two days later, in a telegram to Governor Long, the group begged Long to intervene. Using a state's rights argument, the council implored, "Only the Governor, exercising the inherent right of the sovereign state to maintain the welfare of the state, is in the position to block race mixing at LSUNO. If you want to maintain segregation, it is time to take off the kid gloves and start scrapping. It is time for more than lip service to segregation" ("New LSU Branch Opens with Classes Mixed under Federal Court Edict," 1958).

Gremillion, who had unsuccessfully appealed the preliminary injunction allowing Negroes enter LSUNO, spoke at a rally of the White Citizen's Council, calling for "massive resistance" to integration. His comments were clearly aimed at inciting the group to further action. He entreated the group:

> What can we as citizens do through massive resistance? First, we can vigorously exercise our basic constitutional rights. The right to assemble, the right to organize, the right to protest, your wonderful right of freedom of speech, and above all, we must exercise our right to vote . . . We will have to continue to support and maintain the separate but equal school facilities.
>
> ("New LSU Branch Opens with Classes Mixed under
> Federal Court Edict," 1958)

In the midst of the LSUNO controversy, the LSU Board petitioned State Superintendent Shelby Jackson for a report upon the status of the New Orleans branch of Southern University. The board pressed for early establishment of the new college to relieve the pressure from Negroes for admittance to LSUNO ("SU Branch Plan Talks Are Set," 1958). The attitude prevailed that if Negroes were presented with the "separate but equal" Southern branch college, even in the post-*Brown* desegregation era, they would choose to attend the Negro college instead of LSUNO.

At the joint meeting of the Louisiana State Board of Education (LSBOE) and the LSU Board held on November 1, 1958, Agenda Item 1 contained a status report on Southern branch campus. In the report, Jackson provided details on the college's opening. He elaborated that the college site had been purchased and planning had begun for building a $750,000, three-level modern structure to handle approximately

500 students. In addition, Jackson boasted that construction was scheduled to begin in January or February of 1959 and that final plans for the university included the ultimate construction of 11 buildings on the 22-acre site (State Board of Education and LSU Board of Supervisors, 1958).

Shortly after the meeting, Theo Cangelosi, president of the LSU Board, made an appearance on a New Orleans television program. During the interview, Cangelosi responded to a series of questions concerning the status of Negro students both on the LSU campus in Baton Rouge and the newly opened LSUNO. Although LSU still had no Negro undergraduate students, the total Negro enrollment in the graduate, law, and social welfare schools had declined dramatically from an all-time high of 302 students enrolled in the 1955 summer session following the *Brown* decision to 45 students in the fall semester of 1958. Commenting on the reason for the decline, Cangelosi said,

> It is an interesting thing. We know the reason that it dropped. In 1957, Southern University, which is one of the magnificent state-supported colored universities of this county . . . put in a graduate school. The significant thing is that the very first year that Southern University graduate school was opened, our colored student enrollment dropped from 260 to 99. My personal view, and I think of many thinking people, is that it is evident that colored folks would rather go to school with their own people if the facilities are provided.
>
> ("Segregation in Sports Invalidated by Court," 1958)

Cangelosi further opined that when Southern University opened its New Orleans branch in the following year, Negroes would prefer attending the new facility rather than going to LSUNO ("Segregation in Sports Invalidated by Court," 1958).

RESISTANCE IN THE NEGRO COMMUNITY

The voice of the Negro community remained silent during the initial planning phase for the New Orleans branch of Southern University. Between 1956 and 1958, while the governor, the state legislature, and the SBOE proceeded with plans for the institution, the local Negro community focused most of its efforts on desegregation at the elementary and secondary level, an area of tremendous need. With the successful integration of LSUNO, however, the community turned its attention to the proposed Southern branch. If there were those in the Negro community who supported the idea, they remained silent. However, the voices of opposition, spearheaded by a number of prominent New Orleans Negro organizations, began to rise.

The first sign of opposition came in late September of 1958. The officers and members of the United Clubs, Inc. issued a statement opposing the $750,000 proposed allocation of state funds for the Southern extension. The group, formed in 1954 in an attempt to serve the needs of the Negro community, already sponsored a ball for the United Negro College Fund (UNCF) and their argument attacked the soundness of the plan on two fronts: first, lack of funds and second, segregation ("Rap Proposed SU Extension School Here," 1958).

The organization reasoned that the allocation would cut short the current terms of Louisiana schools by several months because of the lack of sufficient operating funds. On the issue of segregation, the group issued a proclamation. In part, it read,

> It is our whole-hearted conviction that Negroes are in LSUNO and they plan to stay there. The creation of a new building of "our own" will not change the desire of the people of color to exercise their right to attend the tax supported institution of their choice in preference to "one planned exclusively for them." Money was wasted when a separate law school was erected on Southern's campus a few years ago just to keep a few Negroes out of LSU . . . Money will be needed to expand the current site of LSUNO. We recommend the payment of teachers' salaries or the expanding of facilities at LSUNO rather than the waste of funds on maintaining or expanding segregated facilities.
> ("Rap Proposed SU Extension School Here," 1958)

A short time after the United Clubs, Inc. declared opposition, another Negro organization followed suit. In a letter to several civic organizations, the LSBOE, and governor Long, the group also used the fiscal crisis in the state and continued public school segregation as a reason to oppose the Southern branch ("Southern U. Unit Here is Opposed," 1958; "Waste of Money to Build SU Extension Says Study," 1958). "Today, the schools in the state of Louisiana are in a serious financial crisis," the organization asserted; pointing out the needless duplication of the proposed school and the amount of money that would be wasted, declaring, "such a step would be wasteful at the very least even if we had the money to do it with" ("Waste of Money to Build SU Extension Says Study," 1958).

The group's opposition centered on the issue of the state's efforts to perpetuate segregation in Louisiana's postsecondary institutions. The group supported the expansion of institutions on the higher education level, but not on a segregated level ("N.O. Southern U. Branch Opposed," 1958). Keying in on this theme, the group decried

> the needlessness of building an extension of Southern University in New Orleans because Louisiana State University's commuter college is already in existence here. The reason and only reason, admittedly, is to attempt to maintain segregation at the undergraduate level. If this attempt were likely to succeed, then perhaps there would be no need for this letter.
> ("Waste of Money to Build SU Extension Says Study," 1958)

Especially because Negro students had been admitted to LSUNO and were already attending classes there, as well as the fact that Negroes also attended undergraduate colleges in other parts of the state, the letter posed several questions:

> Do you honestly believe that the building of an extension of Southern University will have the effect of keeping Negro students out of LSU in New Orleans? If there is the slightest doubt in your mind, then at the very least, such a proposal should be postponed until the whole segregation-integration picture is clearer.
> ("Waste of Money to Build SU Extension Says Study," 1958)

Next, it begged another question: "Is this a segregation issue or is this a political maneuver and beset with emotional ranting?" ("Waste of Money to Build SU Extension Says

Study," 1958). In an attempt to answer this question and use a logic that might appeal to the letter's recipients, the group made its final argument. The letter concluded,

> Whether one is a segregationist or integrationist or just plain Mr. Citizen, New Orleans, Louisiana, good common sense should prevail. If you believe that in spite of the above, money should be spent on an extension of Southern University . . . then just sit quietly by and watch the farce become a reality. If on the other hand, you believe that good common sense and business prudence dictate that this is not the time to waste educational funds foolishly—then do something about it.
> ("Waste of Money to Build SU Extension Says Study," 1958)

By December 1958, three Negro organizations, the United Clubs, Inc., the Studs Club, Inc., and the Frontiers of America united in their efforts to halt the construction of Southern's extension in New Orleans and reached out to the Negro community for support. Led by the United Clubs, Inc., whose president Dr. Leonard Burns vehemently opposed the new branch, the groups circulated a petition in hopes of gathering 50,000 signatures to oppose the construction scheduled to begin the following month in January 1959 ("N.O. Southern U. Branch Opposed," 1958). Despite these last minute efforts, construction proceeded as scheduled.

CONSTRUCTION OF SUNO BEGINS

Plausibly the driving of the first piling on January 12, 1959, brought a kind of relief to state and local officials as construction finally began on SUNO's $700,000 classroom administration building. The three-story concrete and brick building, constructed by Perrilliat–Bickley Construction Company, was expected to be completed by September 1959, when school opened.

In ceremonies highlighting this historical event, Nash Roberts, SBOE chair, called the beginning of the work on the new Negro college "a great day for education in the state of Louisiana" ("Break Ground for Commuters College," 1959; "Work Is Begun on New College," 1959). He boasted of Pontchartrain Park as "an ideal site for a college, with its golf course, lighted stadium, lagoons and fishing ponds" ("Break Ground for Commuters College," 1959; "Work Is Begun on New College," 1959). Roberts described the area as beautiful and very accessible, and declared the branch to be a better investment of state funds than continuing to spend monies on sleeping and eating facilities on the Baton Rouge campus ("Break Ground for Commuters College," 1959; "Work Is Begun on New College," 1959).

New Orleans Mayor deLesseps S. Morrison, speaking at the ceremonies, hailed the groundbreaking event as evidence of "progress" in the city because of the opening of branches by both Southern University and LSU. He indicated that Pontchartrain Park stood out as a model community and promised that within a year the campus would be fully accessible with the construction of a four-lane drive to the new university branch.

Other state officials attending the ceremonies included State Superintendent of Education Shelby Jackson, SBOE members Alfred E. Roberts of Lake Charles, Eleanore H. Meade of Gremmercy, Joseph J. Davies Jr. of Arabi, Leon Gary of Houma, George T. Madison of Bastrop, Isom J. Guillory of Eunice, and chairman of the building committee, Robert H. Curry of Shreveport. Four Southern University officials also attended the groundbreaking ceremonies, including Dr. Felton G. Clark, president, G. Leon Netterville, business manager, B. A. Little, university auditor, and Dr. Elton C. Harrison, coordinator of instruction ("Break Ground for Commuters College," 1959; "Work Is Begun on New College," 1959).

No one mentioned integration at the dedication. Although there were four senior level administrators at the ceremony from Southern University, their failure to comment might have puzzled onlookers. However, the reason became clear when Little later described the atmosphere at the groundbreaking: "We respectfully broke ground for this school we did not want during the days of segregation. We didn't want SUNO because we didn't think it was necessary at the time" (*Louisiana Weekly*, September 22, 1984). It appeared that no one in the Negro community wanted the university, not even the parent body.

THE PUBLIC RESPONDS

Three days after construction began on the first building for the new campus, an editorial appeared in the *Times Picayune* ("Southern's Branch," 1959). The White-owned newspaper voiced its support for the Southern branch, echoing remarks by officials at the September 13 ceremony. "The city and state can mark the attainment of two important goals within one 12 month span," the editorial stated. "The commuter branch of Louisiana State University was opened last September and the arrival of Southern University in a location particularly well suited for the facility is to be noted with pride and great expectations by a progressing area" ("Southern's Branch," 1959).

Meanwhile, the local Negro press reported reactions of a different kind from within the Negro community. On January 24, 1959, *Louisiana Weekly* boldly declared "Thousands Petition against 'Jim Crow' Commuters College" on its front page, strategically placed next to an article containing registration information for LSUNO. Over 3,500 signatures, representing a "cross-section" of New Orleans, had been gathered through the combined efforts of the United Clubs, Inc., the Studs Club, Inc., and the Frontiers of America, expressing complete opposition to the New Orleans branch of segregated Southern University. Sent to Governor Long the petition asked that construction of the institution be halted. Along with the release of details concerning the petition, the United Clubs, Inc. issued the following statement:

> The many citizens involved are by no means opposed to the promotion of facilities of higher education in the state of Louisiana. But they are diametrically opposed to the use of state funds to build segregated institutions at a time when the entire world has voiced its objection to the denial of the rights' of the individual. The erection of the proposed Southern University branch in New Orleans is defiance to educational progress and the inevitable social change currently in progress. It is a direct insult to more than 200,000 Negroes in New Orleans, Louisiana who do not want a school of "their own."
>
> If thousands of citizens took the time to say we do not want an institution which accommodates Negroes only, then ten times that many say that they will never send their children to this segregated throw back to a period many years past. Southern was taken from New Orleans forty-four years ago as a segregated school.

The Negro would rather attend classes in old buildings at LSUNO than attend classes in new buildings at Southern University in N.O.

("Thousands Petition against 'Jim Crow' Commuters College," 1959)

As the Negro community worked to undermine the new Negro college, the SBOE continued its plans for the establishment of the Southern branch, never identifying the institution as being "across the tracks" from LSUNO. By February, fiscal inequities between the two institutions already surfaced when the SBOE announced plans for "the first phase of a $21 million dollar capital improvements program at LSUNO, compared to the $700,000 appropriation to SUNO [as the institution was now being referred to in the community]" ("Rights Probers Denied Registration Data in Two Northern Parishes of State," 1950; "SU Branch Labeled Unwise Move by UCI," 1959).

News of the plans brought an immediate response from United Clubs, Inc.'s President Dr. Leonard Burns, the most vocal opponent of SUNO in the Negro community. He said,

We are not against education or the expansion of educational facilities, but we are of the opinion that the establishment of special, separate or designated schools for Negroes does not represent the thinking of progressive-minded individuals of both racial groups.

("Rights Probers Denied Registration Data in Two Northern Parishes of State," 1950; "SU Branch Labeled Unwise Move by UCI," 1959)

Dr. Burns further pointed out the disparities in funding between the two institutions: "Since integrated LSUNO is operating now, it is obvious that the 22 acre SUNO to be located near Pontchartrain Park is not only a shoe string project, but a method proposed and designed to keep Negroes away from LSUNO." He further added that a combination of the two institutions in New Orleans would be a major step forward in the establishment of one top-rated state supported institution of higher learning ("Rights Probers Denied Registration Data in Two Northern Parishes of State," 1950; "SU Branch Labeled Unwise Move by UCI," 1959). Dr. Burn's idea of merging the two institutions recurred as a possibility throughout the history and development of SUNO (Cook 1992; "Quit Playing Games . . . Merge SUNO with LSUNO Suggests Urban League," 1969; "SUNO Is Lacking in Quality View," 1969)—although the recognition of Southern University as a separate university system apart from the LSBOE, the LSU system, and the University of Louisiana System dampened this possibility by constitutional fiat (State of Louisiana, 1973).

As the Negro public high schools in the city were preparing their seniors for graduation in May 1959 and decisions were still being made about what college to attend, a new organization took shape, designed to be a sounding board on education for Negro citizens. Through the combined efforts of five of New Orleans' most powerful and prominent Negro organizations, the Citizens Committee for Higher Education for Negroes (CCHEN) was formed. The committee included members of The United Clubs, Inc., The New Orleans Chapter of Frontiers of America, The Studs Club, Inc., The Coordinating Council of Greater New Orleans, and the Urban League of New Orleans ("Inferiority of SU Extension Blasted," 1959).

One immediate goal of the group involved the task to "instill in the minds of both high school graduates of 1959 and their parents the many advantages offered in an integrated publicly supported state extension school, against those to be found in one specifically set up for Negroes" ("Inferiority of SU Extension Blasted," 1959). CCHEN had done their homework. They compiled information for presentation to parents and students and titled it "LSUNO vs. SUNO." In an open letter to high school seniors and their parents, the group clearly laid out their position on higher education for Negroes in Louisiana. CCHEN did not contest "any already established school for higher education for Negroes, but to an extension of segregated facilities." In highlighting one of the major inequities between the LSUNO and SUNO, the letter pointed to the legislative plans to spend more than $21 million on the first phase of LSUNO, located on a 178-acre tract of land on the lakefront, as compared to $750,000 for the first phase of SUNO to be built on a small 22-acre tract of land in the Pontchartrain Park subdivision ("Inferiority of SU Extension Blasted," 1959).

The appeal called upon the emotions of a community still fighting the massive resistance to desegregation by Whites. It read,

> The new Negro does not want to be treated as either a special citizen or a ward of the state, but as another citizen able to take full advantage of all opportunities. Regardless of the howls of the racists, the disciples of hate and their threats, we Negroes must be firm and courageous in this fight for freedom and should not permit our birthrights to be sold on the premise and theory that we are different and want to be left alone. As Americans, as taxpayers, we are supposed to enjoy all the benefits and share the responsibilities this great nation offers.
>
> We are making this appeal to parents to assist their children to make the right turn in the fork in the road of education, which now faces them. One leads to the whole loaf of educational opportunities, while the other leads to the crumbs. We must have the full, whole loaf.
>
> ("Inferiority of SU Extension Blasted" 1959)

The CCHEN stepped up verbal attacks on SUNO in June, warning high school seniors and their parents not to become victims of sales gimmicks relative to facilities of the Southern branch under construction. Dr. George B. Talbert, coordinator of CCHEN, remained as fiercely opposed to SUNO as Burns, and became a spokesman for the opposition forces. Talbert claimed that the White press glamorized SUNO to make it as attractive and appealing as possible to Negroes in New Orleans, as evidenced by a recent photo showing SUNO's first building under construction. In an open appeal to Black citizens of New Orleans that appeared in a local Negro newspaper, Dr. Talbert warned of trouble:

> We are of the opinion that a wholesale enrollment this fall would be a deadly weapon in the hands of segregationists who are already saying that Negroes do not want to integrate; it is just a few of their leaders advocating desegregation. To make use of those facilities would be setting our fight for full citizenship, equal opportunities and human dignity back to the dark ages of hope prior to 1954 . . .
>
> By no stretch of the imagination can we make a comparison of the allocation for the two facilities and sanely conclude that less than one million is equal to $21 million . . . Negroes in general are craving for the luxuries of integrated education,

but many are lacking that which is often necessary to withstand the pressures which often accompany any crusaders into a foreign land, or advocation of "our way of life in the South."

Let cobwebs of protest be spun across the portals of SUNO as concrete signs of rebellion against the extension of segregation, and may the halls and classrooms of LSUNO assume a checkerboard like pattern among its student body. There will never be an understanding between the races if they continue to live apart, work apart and play separately. We must better human relations crushed beyond repair by the many evils race hatred gives birth to.

("Attacks Publicity for Branch College," 1959)

THE LSU BOARD OF SUPERVISORS CONTINUES TO FIGHT INTEGRATION AT LSUNO

With no indication of any change in construction plans for SUNO, and as the Negro community continued to focus its efforts on attacking the institution, few took notice of the fact that the LSU Board had attempted to resegregate LSUNO. Bolstered by the news that SUNO would open in the fall of 1959, the board appealed to the U.S. Fifth Circuit Court of Appeals to upset the ruling of the lower court that resulted in desegregation of LSUNO the previous fall on the grounds that Negroes filing suit failed to exhaust administrative remedies before taking the case to federal court. Again, the board wanted the 80 Negroes attending the university out and to prohibit Negro registration in the future. However, the court handed down a brief opinion that upheld the federal district court order enjoining officials of LSU from keeping Negroes out of LSUNO ("Segregation Made Big Issue in Race for Governor," 1959).

A DEAN IS SELECTED FOR SUNO

As construction continued on the SUNO campus through the summer of 1959, Dr. Clark, president of SU, had yet to name his choice for dean of the new campus. In the first week in August, 1959, Clark tendered the name of Dr. Emmett W. Bashful to the SBOE to fill the position of acting dean of SUNO. Bashful had grown up in Baton Rouge and his father had been a close friend of Dr. Joseph S. Clark, then president of the Baton Rouge campus and father of Felton Clark. Bashful spent his first year in college at Leland University, a private HBCU located in Baker, Louisiana, but left the institution at the beginning of World War II. Immediately after the war, where he served as a decorated officer, Bashful returned to school and completed his undergraduate degree at Southern University. Using the GI Bill, he obtained both his Master's and PhD degrees in political science from the University of Illinois. As a newly minted PhD, Bashful joined the political science department at Florida A&M University in 1948. Summoned to appear before what he called "the Florida Legislative Committee on Un-American Activities" in 1957, Bashful had emerged as an outspoken supporter of desegregation and his activities were monitored by government officials in Florida. Bashful offered summary later, claiming Florida's action "was just a way to get Black folks out of the state." Feeling uncomfortable in Florida, Bashful left Florida and joined Southern's political science department in 1958 (Armfield, 1994).

Clark had not informed Bashful of his plan to make him the dean of SUNO and as Clark presented Bashful's name to the SBOE, Bashful was headed to Mexico with his family for a vacation. Instead of speaking directly to Bashful, Clark asked Dr. Elton C. Harrison, then dean of academic affairs at Southern, to telephone Bashful and tell him of the news. Bashful expressed his reluctance to accept the deanship of SUNO because he had only been at Southern for one year and had no administrative experience to undertake such a Herculean task. But, after further discussion about the embarrassing position for Clark if he refused, Bashful agreed to become dean, for one year (Southern University in New Orleans, 1979).

Within days, Bashful visited New Orleans and examined the campus. What he found discouraged him. His most pressing problem involved assembling a faculty for the fledgling institution. While several persons had been transferred from the Baton Rouge campus, staffing remained incomplete—as did the building, which became his next task since the college was due a September opening. The contractor refused to give a date for the building's occupation. Bashful pressed for a firm date, and the contractors scheduled for a September 21, 1959, completion with the exception that some work would continue on some parts of the building after it opened. More than the problems of an unfinished building and incomplete staff, Bashful also recognized that "a chorus of opposition to the school in the Negro community could be heard" (Bashful, 1987, pp. 2–3).

HELP FROM THE NEGRO COMMUNITY FINALLY ARRIVES—AMIDST CONTROVERSY

There was no office space on the campus to begin the processes essential to opening the school. Bashful needed to set up an office immediately to begin preregistration tasks related to SUNO's opening. When Southern's president, Clark, appealed to the sales manager for Pontchartrain Park Homes for a temporary registration site on Bashful's behalf, he was told that no space was available. Bethany United Methodist Church, located a short distance from the SUNO campus in Pontchartrain Park, emerged as another possibility for a temporary site because it had a room being used as a community center and could easily accommodate SUNO's needs. Dr. G. L. Netterville, vice president of Southern University and a prominent member of the United Methodist Church, called the Reverend Donald Frank, pastor of the church, to ask for his help. Frank brought the request before his church board, which agreed to allow SUNO use of the facility free of charge. The makeshift office had a staff of three; besides Dean Bashful, there were Registrar Herman F. Plunkett and Business Manager Charles E. Burns (Bashful, 1987; Southern University in New Orleans, 1979). Bashful later acknowledged Bethany's contribution: "Historically, we must take note of the fact that SUNO began in New Orleans at Bethany" (Bashful, 1987).

During the third week of August 1959 Nash Roberts, SBOE president, announced that registration activities for SUNO would be held at Bethany on September 3, 4, and 5, with freshmen orientation scheduled for September 8. Classes were set to begin on September 11 and the final date for adding subjects for credit or making changes was September 26. The news that "Jim Crow SUNO" was being given "aid and comfort" by Bethany Methodist Church stunned many in the Negro community and stirred up a different controversy when Resident Bishop Willis J. King received complaints about the use of church facilities by a tax-supported state institution. Bishop King claimed that he had no prior knowledge of the action taken by Bethany and church policy forbade getting involved in controversial issues, particularly those involving the church and state ("Methodist Church 'Used' to Set Up Jim Crow SUNO," 1959).

On the heels of the announcement of SUNO's temporary placement in Bethany, CCHEN launched a last-ditch effort campaign to steer students toward LSUNO by distributing LSUNO pamphlets in the Negro community. The committee got an additional reprieve when SUNO failed to open as planned and registration became delayed until September 21 and 22, with the first day of classes rescheduled for September 23 ("Methodist Church 'Used' to Set Up Jim Crow SUNO," 1959). But, to no avail—SUNO opened.

SUNO OPENS

When registration began on September 21, 1959, in the first building constructed on the new campus, no welcome mats or long lines greeted students waiting to register. In fact, "Nothing much is being offered" bubbled up as the typical appraisal from students ("Methodist Church 'Used' to Set Up Jim Crow SUNO," 1959). One hundred and fifty eight students registered for admission for the fall of 1959, including transfer students. The initial student body was made up of students from New Orleans as well as seven other nearby parishes, which included Jefferson, Plaquemines, St. Bernard, St. Charles, St. James, St. John, and St. Tammany. At least 88% of the applicants and enrollees came from New Orleans (Southern University at New Orleans 1959–1999: A Wilderness Journey, p. 9). Like LSUNO, only freshmen were admitted the first year. An additional class would be added each year thereafter.

While it had been proposed that only ten basic entry-level courses would be offered, the admission of transfer students required courses beyond the freshman level and presented a problem. Nevertheless, students chose from only 15 courses offered: Theory and Musicianship 111, Piano 111, University Choir 111, English 111, French 111, Spanish 111, Geography 110, American History 110, and Typewriting 211. The administrative structure involved four academic divisions, which also constituted the majors that were available: Humanities, Science, Social Science, and Commerce (Southern University at New Orleans 1959–1999: A Wilderness Journey). Thus, on September 23, 1959, a fully segregated SU at New Orleans began its first day of classes with a freshmen class of 158 students, one building, a faculty of 15, and a staff of 13. It was the first of only two public four-year HBCUs created after the *Brown v. Board of Education* decision (the other being the University of the Virgin Islands in 1962).

SUNO—A CONCLUSION AND KATRINA POST SCRIPT

On June 7, 1892, Homer Plessy challenged state segregation statutes by boarding a White-only car in the city of New Orleans (Brown-Scott, 2000) and unwillingly abetted in the Supreme Court decision that legitimized the creation of separate facilities and schools for Blacks and Whites. Although in the 1930s Southern social scientists presaged the folly and failed economy of building a dual system of education (Nixon,

1934), Louisiana persisted and laid the groundwork for continued segregation in bricks and mortar. Now, over one century after *Plessy v. Ferguson* (1896) and a half-century after *Brown v. Board of Education* (1954), reconsideration of the social and public policy context that gave rise to two separate and unequal public universities just a few miles apart in one city begs reflection. Certainly, the widespread devastation following Hurricane Katrina (August 29, 2005) and massive flooding from the ensuing levee failures ("Scattered Lives," 2005), brings a new chapter to the storied histories of each New Orleans university, and SUNO in particular.

SUNO faces many fits and struggles in reviving itself in a post-Katrina city and its recovery pace has lagged severely behind other New Orleans universities (Selingo, 2005). Hit harder than any college in the region with all of its buildings "under 10 to 12 feet of water for weeks" (Mangan, 2007), SUNO reopened in the spring 2006 semester using an alternate location (Mangan, 2005; Walters, 2005). Two years later, the university operates from its North Campus out of "45 modular trailers" (Blum, "N.O. colleges censured," 2007) with only the promise of "state officials to start making repairs soon" (Mangan, 2007).

Damage estimates for SUNO peaked at $350 million and the prognosis advised a complete rebuilding (Schuman, 2005). Figuratively, the university has existed on life support since 2005, with an intermittent flow of federal, state, and international dollars (Mangan, "Colleges to Share," 2006). To reduce expenses quickly the university's Board of Supervisors laid off 40% of its faculty, lopped off 19 programs, and strategically expanded online offerings in targeted degree programs such as criminal justice and museum studies (Mangan, "Still without a Campus," 2006; Howard, 2006). Despite a few negative audit findings, the SU system net assets increased in value modestly from 2006 to 2007 (Blum, "SU System's Finances Better," 2007). The prospects of SUNO depend largely upon the ability of the university to increase student enrollment, which hovers at 65% of its pre-Katrina figure (Gyan, 2007).

Providing safe and affordable housing stands out as a key aspect of the city's repopulation and is necessary for fueling the university's enrollment growth, a fact not lost upon SUNO staff and students who have worn the emotional strain of "living and working out of trailers 24/7" in a completely "devastated neighborhood," according to Louisiana's Higher Education Commissioner E. Joseph Savoie and Chancellor Victor Ukpolo (Mangan, 2007). To the positive, SUNO secured a $44 million low-interest loan from the federally funded HBCU Capital Financing Program to construct the first student housing for the campus, a project that will provide 650 new and much-needed beds (Blum, "SUNO scaling down its dormitory project," 2007). However, it has also received scrutiny along with four other New Orleans institutions when it received a censure in May 2007 by the Association of American University Professors (AAUP) for faculty mistreatment in the hurricane's aftermath (Gravois, 2007; Blum, "N.O. colleges censured," 2007). Though SUNO defended itself, claiming that the AAUP demonstrated "insensitivity" to the university's extreme circumstances, the censure may have some negative consequence upon the task of attracting new faculty (Blum, "N.O. colleges censured," 2007).

This history underscores the fact that SUNO exists and functions with the context of a race-stratified society. As a state-supported minority-serving institution with open enrollment (Fischer, 2007; Mangan, "Still without a Campus," 2006), the university's existence, purpose, and value have been questioned periodically throughout its history—first by the Black community consigned to support and attend it; periodically by Louisiana's legislature; and in the storm's aftermath, by the educational governance agencies charged with accountability, resource distribution, and reform (Fischer, 2006; Mangan, "Still without a Campus," 2006). Founded in proximity to UNO—a state university with a mission to serve the metropolitan area whose existence and function has not received similar question or scrutiny, SUNO stands as a living legacy to America's privilege and power divide and its history of resource disparity and uneven support is but part and parcel of its status and larger mission.

However, SUNO is also a young university, at 51 years of age, and its growing pains have been very public and part of recent memory more so than the establishment and early development of the colonial colleges, for example. Higher education history demonstrates that universities need financial and capital resources to grow and thrive, and arguably SUNO needs these now, if not more than ever. However, other resources, including individual and community energies, nurture universities in ways rarely and uneasily accounted for in the historical record. Despite tremendous obstacles and economic disparities particularly in their early years, many universities evolve and transform over time, moving from a beleaguered to beloved status. Yet on this uncertain continuum the number of years, the exact formula required for such institutional transformations, and who judges remains remarkably unclear. Undoubtedly, *if* institutional resilience, individual persistence, and collective determination are decisive factors in identifying whether an institution is beleaguered or beloved, then SUNO cannot just be beleaguered as it meets the challenges of a new era, it is certainly beloved too.

REFERENCES

Armfield, F. 1994. Behind the veil with Emmett Bashful. Personal Interview with Emmett Bashful. Duke Library, June 24.

Attacks publicity for branch college. 1959. *Louisiana Weekly*, June 1.

Bartley, N. 1969. *The rise of massive resistance: Race and politics in the south during the 1950s.* Baton Rouge, LA: Louisiana State University Press.

Bashful, E. 1987. Reach for the stars. Speech given to faculty, staff, students, alumni and friends at Southern University in New Orleans. May 5.

Blum, J. 2007. N.O. colleges censured: Schools taken to task for treatment of faculty. *Advocate*, June 12, A10.

———. 2007. SUNO scaling down its dormitory project. *Advocate*, June 13, A14.

———. 2007. SU system's finances better, legislative audit concludes. *Advocate*, June 19, B1.

Break ground for commuters college. 1959. *Louisiana Weekly*, January 17.

Brown-Scott W. 2000. Plessy v. Ferguson (1896). In *Civil rights in the United States,* Vol. 2, ed. W. E. Martin and P. Sullivan. New York: Macmillan Reference USA.

Brown v. Board of Education of Topeka, 347 U.S. 483 (1954).

Clark, F. 1958. The development and present status of publicly-supported higher education for negroes. *Journal for Negro Education* 27(3): 221–32.

College segregation-desegregation issue revived by new court action. 1957, February. *Southern School News*.

Colleges given house unit okay. 1954. *Times Picayune*, June 2, A1, A5.

Committee of the Faculty of the College of Education, Louisiana State University. 1969. *The development of public education in Louisiana*. Baton Rouge, LA: Louisiana State University.

Cook, R. 1992. Pro and con of SUNO/UNO merger. *Driftwood*, October 22.

Dream house. 2003. *Times Picayune*, August 17.

Fenton, J. H. 1957. The Negro voter in Louisiana. *Journal for Negro Education* 26(3): 319–28.

Fischer, K. 2007. A historically Black college takes a hands-on approach to student success. *Chronicle of Higher Education*, March 23, A21.

Fischer, K. 2006. Finding promise in pain. *Chronicle of Higher Education*, March 10, A21.

Gravois, J. 2007. AAUP carries through with plan to censure 4 New Orleans universities. *Chronicle of Higher Education*, June 22, A12.

Gyan, J. 2007. Obstacles don't deter repopulation. *Advocate*, June 22, A13.

Haskew, L. 1968. Impact of the southern regional education board in its first twenty years. *The future South and higher education*. Atlanta: South Regional Education Board.

Henley et al. v. Board of Supervisors of LSU, C.A. 2105. Unreported.

Howard, J. 2006. Southern plans to cut 19 programs. *Chronicle of Higher Education*, January 6, A40.

Inferiority of SU extension blasted. 1959. *Louisiana Weekly*, May 16.

Johnnson, L., D. Cobb-Roberts, and B. Shircliffe. 2007. African Americans and the struggle for opportunity in Florida public higher education. *History of Education Quarterly* 47(3): 328–58.

Kurtz, M. 2001. Earl Long's political relations with the city of New Orleans: 1948–1960. In *The Louisiana bicentennial series in Louisiana history, Vol. 8. The age of the Longs: Louisiana 1928–1960*, series ed. E. F. Haas, 464–74. Lafayette, LA: Center for Louisiana Studies.

Kurtz, M., and M. Peoples. 1990. *Earl K. Long: The saga of uncle earl and Louisiana politics*. Baton Rouge, LA: Louisiana State University Press.

Legislature investigates pro-integration feeling in nine colleges. 1958, July. *Southern School News*.

Louisiana. 1954, October 1. *Southern School News*.

Louisiana governor pledged to ban schools if courts act. 1956, February 10. *Southern School News*.

LSU board "taunts" 53 Negro students. 1958. *Louisiana Weekly*, September 20, 1.

Mangan, K. 2005. 2 colleges hit by Katrina cut their staffs by over half. *Chronicle of Higher Education*, November 11, A1.

———. 2006. Still without a campus, Southern U. at New Orleans struggles to stay in business. *Chronicle of Higher Education*, May 26, A31.

———. 2006. Colleges to share Katrina money. *Chronicle of Higher Education*, September 8, A27.

———. 2007. Some New Orleans colleges predict bigger enrollments this fall. *Chronicle of Higher Education*, May 18, A32.

Methodist church "used" to set up Jim Crow SUNO. 1959. *Louisiana Weekly*, August 29.

New LSU branch opens with classes mixed under federal court edict. 1958, October. *Southern School News*.

Nixon, H. 1934. Colleges and universities. In *Culture in the south*, ed. W. T. Couch, 229–47. Chapel Hill, NC: University of North Carolina Press.

N.O. Southern U. branch opposed. 1958. *Times Picayune,* December 7, A8.

100 Negroes registered in Louisiana colleges under courts' injunction. 1957, March. *Southern School News.*

One student is expelled in incidents. 1958. *Louisiana Weekly,* September 28.

Option steps for Pontchartrain site for college branch okayed. 1957. *Times Picayune,* May 12, B7, B8.

Orfield, G. 1969. *The reconstruction of southern education: The schools and the 1964 Civil Rights Act.* New York: John Wiley.

Payne v. Board of Supervisors of LSU, C.A. Unreported.

Plessy v. Ferguson, 163 U.S. 537 (1896).

Quit playing games . . . merge SUNO with LSUNO suggests urban league. 1969. *Louisiana Weekly,* May 24.

Rap proposed SU extension school here. 1958. *Louisiana Weekly,* September 27.

Rights probers denied registration data in two northern parishes of state. 1950, April. *Southern School News.*

Scattered lives. 2005. *Chronicle of Higher Education,* September 16, A10–17.

School issue sharpened in Louisiana as new year begins. 1955, September. *Southern School News.*

Schuman, J. 2005. Southern U. at New Orleans may have to rebuild from scratch. *Chronicle of Higher Education,* September 23, A16.

Segregation in sports invalidated by court. 1958, December. *Southern School News.*

Segregation made big issue in race for governor. 1959, May. *Southern School News.*

Selingo, J. 2005. Tulane U. sets the pace for recovery. *Chronicle of Higher Education,* May 26, A30.

Southern U. unit here is opposed. 1958. *Times Picayune,* September 30, A3, A4.

Southern's Branch. 1959. *Times Picayune,* September 15, B1.

Southern University at New Orleans 1959–1999: A Wilderness Journey.

Southern University in New Orleans. 1979. *20th Anniversary Brochure (1959–1979).* New Orleans, LA: Southern University.

State Board of Education and LSU Board of Supervisors. 1958, November 1. *Minutes of the Joint Meeting.*

State Department of Education. 1957, January 25. *Official Proceedings of the State Board of Education, Bulletin No. 840.* Baton Rouge, LA: State Board of Education.

———. 1957, May 3. *Official Proceedings of the State Board of Education, Bulletin No. 842.* Baton Rouge, LA: State Board of Education.

State of Louisiana. 1973. *Minutes from the Constitutional Convention.*

State Department of Louisiana. 1957, June 14. *Official Proceedings of the State Board of Education, Bulletin No. 846.* Baton Rouge, LA: State Board of Education.

State of Louisiana. 1956. *Acts of the legislature (Extraordinary Session of 1956).* Baton Rouge, LA: Thos. J. Moran's.

State of Louisiana. 1956. *Official journal of the proceedings of the state of Louisiana at the twenty-seventh extraordinary session of the legislature.* Baton Rouge, LA: Thos. J. Moran's.

State prepared to acquire site. 1957. *Times Picayune,* July 18, A7, A8.

SU branch labeled unwise move by UCI. 1959. *Louisiana Weekly,* February 28.

SU branch plan talks are set. 1958. *Times Picayune,* November 1, B5, B6.

Sugg, R., and G. Jones. 1951. *The southern regional education board: Ten years of regional cooperation in education.* Baton Rouge, LA: Louisiana State University.

SUNO is lacking in quality view. 1969. *Times Picayune,* May 15, A1, A2.

Tax backed colleges open doors. 1956, March. *Southern School News.*

Thompson, C. 1949. Why Negroes are opposed to segregated regional schools. *Journal of Negro Education* 18: 1–8.

Thousands petition against "Jim Crow" commuters college. 1959. *Louisiana Weekly,* January 24.

Uneventful Louisiana college integration involves 160–175. 1957, November. *Southern School News.*

University of New Orleans. n.d. *UNO history.* Retrieved August 31, 2007, from http://www. uno.edu/history.

Vander Zanden, J. 1958. *The southern White resistance: Race and politics in the south during the 1950s.* Doctoral dissertation, University of North Carolina, Chapel Hill, 1958.

———. 1959. Resistance and social movements. *Social Forces* 31: 312–5.

———. 1962. Accommodation to undesired change: The case of the south. *Journal of Negro Education* 31: 30–5.

Wallenstein, P. 1999. Black southerners and non-Black universities: Desegregating higher education 1935–1967. *History of Higher Education Annual* 19: 121–48.

Walters, A. 2005. Southern U. to reopen soon. *Chronicle of Higher Education,* November 25, A32.

Waste of money to build SU extension says study. 1958. *Louisiana Weekly,* October 4.

Wells, A. 2002. From ideas to institutions: Southern scholars and emerging universities in the South, circa 1920–1950. *Dissertation Abstracts International* 62(06): 2053A. (UMI No. 3018928)

———. 2004. Mischief-making on the eve of *Brown v. Board of Education:* The origins and early controversies of the southern regional education board. Unpublished paper presented at the American Educational Research Association, San Diego, CA.

Wilson v. Board of Supervisors, 92 F Supp. 986 (1950).

Work is begun on new college. 1959. *Times Picayune,* January 13, B7, B8.

Young Men's Business Club of New Orleans, Education Bureau. 1954, March 31. *Resolution.*

CHAPTER 9

THE IMPACT OF A FEDERAL CIVIL RIGHTS INVESTIGATION ON CENTRAL STATE UNIVERSITY OF OHIO, 1981–2005

KRISTEN SAFIER

With the 1954 *Brown v. Board of Education* decision that public schools could not be racially segregated, and the Supreme Court's 1955 instruction that this mandate be implemented "with all deliberate speed," many began to question whether the desegregation decree would lead to the merger of public Black colleges and universities into historically White institutions (HWIs). Federal civil rights legislation, such as Title VI of the 1964 Civil Rights Act, similarly seemed to point toward the potential closures of historically Black colleges and universities (HBCUs). In Ohio, the United States Department of Education's (US DOE) Office for Civil Rights (OCR) concluded in a 1981 report that the state had failed to support its only HBCU, Central State University (CSU), adequately when compared with HWIs. Nevertheless, this question was not squarely addressed by the Supreme Court until its 1992 consideration of *United States v. Fordice*. The *Fordice* decision neither required nor prohibited HBCU closure as a necessary component for integration, leaving the question open for policymakers.

Despite predictions of HBCU closures, however, these institutions continue to graduate nearly one-third of all African Americans receiving college degrees, as well as a majority of Black professionals in fields such as dentistry, medicine, and engineering (Williams, 2004). In Ohio, CSU has remained open despite a recent state investigation and intensive oversight of school finances. The 1981 OCR investigation continued until 1998, overlapping with the state's financial oversight in the mid-1990s.

Ironically, while many predicted HBCU closures in response to desegregation jurisprudence and civil rights legislation, the federal investigation was used by Central State
supporters arguing for the school's continued existence and increased support. Indeed,
Central State officials and supporters allowed the federal investigation to frame the
funding debate that surrounded the school in the 1990s.

This chapter examines the impact of the federal investigation on the relationship
between CSU and the state of Ohio. A brief history of the development of Central
State is given, with particular attention to the role that state funding played in the formation and early growth of the university. Next, the chapter examines civil rights legislation and case law relevant to Central State's status as a publicly supported HBCU,
the only one in Ohio. Central State's more recent history is discussed next, as the
chapter tracks the federal civil rights investigation, as well as state inquiries into the
university's finances, from 1981 through present time. Despite the prediction that
Brown and its progeny could signal the death knell for HBCUs, I argue that Central
State was able to negotiate its continued existence by relying in part on the OCR
investigation. Finally, the chapter concludes with a consideration of whether Central
State's experience could be instructive to other publicly supported HBCUs, or
whether it is an anomaly.

In examining this issue, I use traditional research sources, such as scholarly books
and journal articles on HBCUs. Despite Central State's status as Ohio's only public
Black college, however, the university has received little attention in the scholarly literature on HBCUs. At the recommendation of Sheila Darrow, archivist at CSU,
Lathardus Goggins's (1987) *Central State University: The First One Hundred Years,
1887–1987* served as the primary resource for Central State's early history. Goggins, a
graduate of CSU, chronicled the school's development in unique detail. For more
recent information, I reviewed scholarly and general news sources, including *Chronicle
for Higher Education, Black Issues in Higher Education*, and *Gold Torch*, the CSU student newspaper. State budget appropriations and student information data from the
Ohio Board of Regents (OBR) and the Center for Higher Education at Illinois State
University were consulted, as well as available materials from the Ohio inspector general on the state's investigations into Central State's finances. Because Central State has
requested that the federal government reopen its investigation into Ohio's treatment of
the university, however, little documentation on the federal investigation is currently
available. While news reports and government data were useful in reconstructing a
timeline for Central State, there have been very few analyses of the significance of the
school's history, the reasons for its unique history, and the role that the school has
played in Ohio higher education. Through this chapter, I hope to bring renewed and
worthwhile scholarly attention to CSU. Indeed, a study of its history and its contemporary experiences in the post-*Fordice* legal landscape may illuminate previously overlooked issues in higher education.

CENTRAL STATE UNIVERSITY: ITS FORMATION AND GROWTH[1]

The story of CSU began with the development of Wilberforce University in the mid-
1800s. Ohio African University opened on the site that would become Wilberforce
University in 1856, through the work of the African Methodist Episcopal (AME)

Church. From its inception, the school admitted both free and enslaved Black men and women (Williams, 2004). Although decreases in enrollment and financial support during the Civil War prompted the school to close in 1862, an AME bishop reopened it a year later by moving Union Seminary, a Black college and seminary outside of Columbus, Ohio, to the Wilberforce site and calling it Wilberforce University (Goggins, 1987). The Reverend Daniel Alexander Payne took the helm of the university and became the country's first Black college president (Williams, 2004). After arsonists' destruction of the university's main building in April 1865, the campus and curriculum were restructured (Goggins, 1987). Preparatory and collegiate divisions operated, and the university graduated ministers and teachers beginning in the early 1870s. Because university funding was dependent upon the generosity and availability of benefactors, Wilberforce lobbied the state for funding for a separate division of the university offering vocational and teacher training.

In 1887, Ohio authorized the use of public funds for the development and maintenance of that separate division, the Combined Normal and Industrial Department at Wilberforce University. Because Wilberforce was a church-supported university, the department was governed by a separate board of trustees, the majority of whom were appointed by the governor. No explicit limitations were placed on enrollment by race, but "it was generally believed that only Blacks would wish to attend an institution connected with a Black university and located in an overwhelmingly Black community" (Goggins, 1987, p. 6). Interestingly, in 1887, Ohio both provided state funding for a separate normal school at the church-affiliated university and repealed race segregation in public schools. Indeed, by the late 1800s, most of the state's universities were open to Black students, although few enrolled.

The conflict between the goal of racial integration and the development of Wilberforce as a Black college became evident when the school sought federal Morrill Act funds in 1891. When an Ohio state senator introduced a bill to grant all federal funds to The Ohio State University, a university with an overwhelmingly White student body, Wilberforce President Samuel Mitchell requested an equal division of those funds between Ohio State and the state-supported Combined Normal and Industrial Department at Wilberforce. This debate split public opinion, particularly within the Black community, and the Ohio legislature. In April 1891, former President Rutherford B. Hayes and state officials brokered a compromise in which Ohio State would receive the Morrill Act funds and the state would increase its appropriations for Wilberforce. Indeed, Wilberforce's 1889 grant of $2,000 was raised to $16,000 in 1892, with further increases in 1894 and 1896.

The consequences for Wilberforce, however, were greater than expected. Increased state oversight came with the increased state funds, the Combined Normal and Industrial Department surpassed the privately funded Wilberforce divisions in size and enrollment, and control shifted from university officials to university trustees. Ultimately, increasing state aid divided the state-funded department from the privately funded university such that the privately supported divisions were left in a similarly difficult financial position to that of the pre-department era. With this widening division, the church-supported portions of the school faced greater financial struggles; by 1930, the state paid the private division tuition funds for students enrolled in general education courses through that division. Nevertheless, Wilberforce persevered, thriving in the early

1940s with the development of master's degree programs, coursework in African culture and civilization, and an influx of $1,000,000 in state funding for building renovation and faculty development. In 1944, Wilberforce became the sixth institution in the Inter-University Council of Ohio and received accreditation.

The issue of control continued: in 1945, Ohio Senate Bill 293 proposed giving the state the right to hire and fire the university's president. This brought issues between the AME church and the state of Ohio to the forefront, and a Joint Executive Committee was formed with three trustees from the state board of trustees and three from the church board of trustees. Ultimately, the two boards split when the church board of trustees voted to dismiss Wilberforce President Charles Wesley and the state board voted to support Wesley's move to support the state-supported college from Wilberforce University. The church filed for an injunction against President Wesley from using any of the university's property or supplies, which an Ohio court rejected on the argument that the state had a right to operate a university that it had supported with public funds. Instead, the court allowed the university to separate into two institutions, one private and church-supported and the other publicly funded. In 1951, the Ohio legislature, over the governor's veto, passed legislation to rename the state-supported division Central State College, to operate as an Ohio public college.

Between 1951 and 1965, when Central State reached university status, the school underwent enormous growth. Enrollment increased by 276%, from 790 in 1951 to 2,241 in 1965. The faculty grew from 89 to 101 members, with a larger number of PhD graduates teaching. Across-the-board salary adjustments were made to bring Central State salaries in line with those of the other Ohio public universities. The school expanded to offer graduate degrees in the early 1960s. Its real property holdings nearly tripled between 1947 and 1957, and it renovated 14 and added 12 buildings to the campus by 1964. During this period, federal and state appropriations for higher education also increased dramatically. The federal higher education budget grew 300% between FY1951 and FY1965; the state's capital outlay for higher education increased by 6,000% in that same period (although five were added to the list of state-supported institutions during this time as well). Central State's portion of these funds was small: $117,500 of $2,465,200 in the 1951–53 biennium, and $3,922,000 of $164,620,466 in 1965–67, representing a 3,300% increase. Although state appropriations for operating expenses rose 332% in this period, this actually represented a more than 20% decrease in per-student funding because of increasing enrollment. For Central State, per-student funding declined approximately 1.5% per year from 1951 to 1966. Further, Central State tended to enroll greater numbers of economically disadvantaged students than Ohio's other public colleges, making tuition increases particularly difficult. The school remained careful to maintain its status as a racially integrated university, particularly in response to a 1965 investigation into financial mismanagement and alleged de facto segregation. Indeed, in the mid-1960s, about one-third of the faculty and one-fifth of the student body were White.

During this period of growth, however, a number of official decisions were made that troubled Central State officials. In FY1966, the state began appropriating funds for the development of a new university, which would become Wright State University, only 11 miles from Central State's campus. By FY1968, appropriations for Wright State exceeded those for Central State; by FY1971, Wright State received more than twice the amount that Central State did from the state of Ohio (Center for Higher Education, 2005). Today, Wright State receives nearly five times as much state support as Central State (Center for Higher Education, 2005). In 1971, Central State lost its accreditation for graduate degree programs (Goggins, 1987). University President Lewis Jackson noted, "[The loss of accreditation] leaves some students to wonder why CSU is the only state university without a graduate program. It does not present an image of achievement such as exists elsewhere and may indicate that certain kinds of people are less qualified for graduate studies" (Goggins, 1987, p. 85). In 1974, a tornado devastated the campus, killing 4 people and damaging 13 of the campus' 14 main buildings. The state-funded renovations were, for many Central State officials, long overdue prior to the tornado damage. Nevertheless, by the early 1980s, Central State's financial difficulties had worsened. Further, in 1981, the United States Department of Education OCR investigated Central State and five other publicly supported Ohio colleges. The OCR's conclusion that Ohio failed to support Central State adequately will be explored in Part IV of this chapter. Part III will provide the background for this conclusion by reviewing the legislative and judicial protections against educational segregation that underlie the OCR report.

CIVIL RIGHTS PROTECTIONS AGAINST EDUCATIONAL SEGREGATION

In many states, particularly in the South, Black colleges operated under a system of de jure segregation. Mississippi, for example, operated five public universities exclusively for White students and three public universities exclusively for Black students from the mid-nineteenth to the mid-twentieth century (*United States v. Fordice*, 1992). Indeed, the 1896 case of *Plessy v. Ferguson* explicitly held that separate but equal facilities for persons of different races were constitutionally permissible. The *Plessy* Court pointed to school segregation as proof of the constitutionality of the separate but equal regime. Three years later, the court expressly approved racial segregation of public schools in a case upholding a Georgia school board's decision to turn a county's only Black high school into a White elementary school (*Cummings v. Richmond County Board of Education*, 1899). There, the court rejected the Black plaintiff's motion to enjoin the operation of White high schools, asserting that education was the province of the state. In Ohio, however, the situation was different. In 1887, the state both began funding a department at historically Black Wilberforce University and outlawed racial segregation in schools. Wilberforce, and later Central State, did not operate under a system of de jure segregation then. Indeed, although criticisms of de facto segregation have been levied at Central State, it has historically enjoyed a better record of racial diversity than Ohio's HWIs.

Because of Southern de jure segregation, however, groups, such as the National Association for the Advancement of Colored People (NAACP), developed legal strategies to dismantle segregation through the courts in the mid-twentieth century. Indeed, in most cases, although Black colleges fought to provide Blacks with educational opportunities, years of underfunding and neglect left Black students with fewer options than their White counterparts. In many states, the professional opportunities available to educated Blacks were dictated by the program offerings of publicly supported Black colleges.

The first case in a series of actions brought by the NAACP to challenge segregation in higher education and lay the foundation for *Brown v. Board of Education* held that the state of Missouri must permit a Black applicant to attend an all-White law school or provide a separate but equal in-state institution (*Gaines v. Canada, Registrar of the University of Missouri*, 1938). Until 1938, however, when the case reached the Supreme Court level, Missouri simply did not provide legal education to Blacks. Even after such cases, state resistance to Black education continued. When the state of Oklahoma was ordered to provide a legal education to a Black student, it roped off an area in the state capitol building for her legal training rather than permit her admittance into classes with White students (Gasman, 2004). Texas, Louisiana, North Carolina, and Florida also created separate law schools after lawsuits charged each state with denying Blacks access to legal education (Samuels, 2004).

In an effort to stave off racial integration in public schools, Southern states directed additional funds to Black schools in hopes of destroying the NAACP argument that the segregated education system failed to provide equal resources and opportunities (Samuels, 2004). This effort led the NAACP to reframe its argument as one that assumed all segregated educational institutions to be necessarily unequal (Samuels, 2004). This was the argument that was ultimately upheld by the Supreme Court in *Brown v. Board of Education* (1954). The court concluded that racial segregation was injurious to Black children and denied them equal educational opportunities, even if separate schools were equal. Nevertheless, Albert Samuels (2004, p. 55) has noted that this "strategy left no room for Black colleges once the goal of obtaining legal access to White institutions had been achieved." Many activists involved in the development of this approach argued that the Southern strategy of attempting to equalize funding in the face of desegregation litigation led to enhancements at Black colleges that were wasteful duplications and would halt the goal of integration into White colleges (Samuels, 2004). As the Supreme Court began to reject the newly developed programs for Black students, "Black plaintiffs found themselves in the ironic position of chastising HBCUs for all of their academic shortcomings while at the same time opposing the very remedial measures that would have the effect of improving the quality of these institutions" (Samuels, 2004, p. 57). Nevertheless, although most Black primary and secondary schools closed or merged with White schools after *Brown*, Black colleges remained open (Taylor and Olswang, 1999).

The Civil Rights Act of 1964 enacted federal anti-discrimination laws and created agencies for the enforcement of these laws. Title VI outlawed racial discrimination in schools, a mandate that the OCR was authorized to enforce. Within five years of its passage, the office notified several states that they had violated Title VI by failing "to disassemble racially segregated systems of higher education" (Taylor and Olswang, 1999, p. 75). In *Adams v. Richardson* (1973), a federal appellate court initially upheld the NAACP's claim that the federal government had failed to enforce Title VI against state higher education systems. In 1973, the court ordered that the federal government oversee the plans of 19 states to improve their support for public HBCUs. By 1987, however, the litigation had grown "to colossal, unmanageable proportions," and the plaintiff's claims were dismissed for lack of standing to sue (*Women's Equity Action League v. CAVA205*, 1990). Litigation continued in several other states. In 1992, the United States Supreme Court considered a desegregation challenge to Mississippi's system of higher education in *United States v. Fordice*.

From the mid-nineteenth to the mid-twentieth century, Mississippi had developed a dual public university system with five universities for White students and three for Black students. Even after the 1955 *Brown* II order requiring desegregation "with all deliberate speed," de facto segregation at the state's public colleges continued for decades and with an insufficient state response. The *Fordice* Court first held that race-neutral admissions policies alone would not suffice to demonstrate that the prior dual system had been dismantled:

> If the State perpetuates polices and practices traceable to its prior system that continue to have segregative effects—whether by influencing student enrollment decisions or by fostering segregation in other facets of the university system—and such policies are without sound educational justification and can be practicably eliminated, the State has not satisfied its burden of proving that it has dismantled its prior system.
>
> (*Fordice*, 1992, p. 731)

Using this test, the court concluded that "several surviving aspects of Mississippi's prior dual system . . . are constitutionally suspect," because they "substantially restrict a person's choice of which institution to enter and . . . contribute to the racial identifiability of the eight public universities" (*Fordice*, 1992, p. 733).

The court focused on four specific policies: admission standards, program duplication, institutional missions, and continued operation of all universities. First, higher American College Test (ACT) scores were required for automatic entrance into the HWIs than the HBCUs, despite the similar missions of those schools and without educational justification. The average ACT score for White applicants was three points higher over the required score for automatic entrance into the HWIs; the average score for Black applicants, however, was eight points below the automatic entrance score. Second, the court pointed to a lower court finding that 34.6% of undergraduate and 90% of graduate programs at HBCUs were unnecessarily duplicated at HWIs. Because the "separate but equal" requirement of the *Plessy* era would have necessitated the development of similar programs at both Black and White colleges, the court viewed continuing programmatic similarities with particular suspicion. Third, the court noted that three of the five HWIs had been the state's "flagship institutions" during the period of segregation, receiving the highest level of state funding and developing the most advanced and the widest range of curricular offerings. In 1981, those three institutions were deemed "comprehensive" universities with the largest variety of programs and more graduate-level offerings, while the remaining two HWIs and two of the three HBCUs were designated "regional" universities with fewer programs and a focus on undergraduate education. The third HBCU was classified as an "urban" university, with a mission specific to its urban location. Combined with the differing admissions standards but unnecessary program duplication, the court concluded that the mission assignments would likely interfere with student choice and perpetuate racial segregation. Finally, the court addressed the state's decision to continue operation of all eight universities, despite the geographic proximity of some of the schools. One HBCU was only 35 miles from an HWI; another was only 20 miles from an HWI. Although the court noted that "[e]limination of program duplication and revision of admissions criteria may make institutional closure unnecessary" (*Fordice*, 1992, p. 742), it ordered the lower court to consider this option, within the constitutional framework the Supreme Court delineated, on remand.

In *Fordice*, then, the court did not answer the question of HBCU closure. The justices themselves disagreed over whether closure was warranted. In her concurrence, Justice Sandra D. O'Connor seemed to favor closure of some of the institutions, while Justice Clarence Thomas's concurrence sought to clarify that the closure of HBCUs should not follow. Justice Antonin Scalia, concurring in part and dissenting in part, read the majority opinion as "designed to achieve . . . the elimination of predominantly Black institutions" (*Fordice*, 1992, p. 760). The Fordice parties faced a similar disagreement. Although the private petitioners who originated the lawsuit sought additional funds for Black colleges to counter past discrimination, the Justice Department opposed that remedy when it took over the case (Jaschik, 1991). Instead, the Justice Department brief argued that such funds would "have the perverse effect of encouraging students to attend a school where, other things now being more nearly equal, their own race predominates" (Jaschik, 1991). Commentators, then, questioned whether *Fordice* would require the elimination of HBCUs, an argument for which support lessened when the OCR released a 1994 statement reaffirming its commitment to HBCUs and promising strict scrutiny of any state proposals to close them (Taylor and Olswang, 1999). For Ohio's CSU, it may have been the threat of federal Title VI enforcement that allowed the school to stay open over dwindling state support.

OVERLAPPING INQUIRIES: FEDERAL AND STATE INVESTIGATIONS INTO CENTRAL STATE

Central State University, like its predecessor Wilberforce University, has faced the possibility of closure at several points in its history. Most recently, some Ohio legislators supported the university's closure in the mid-1990s when the school was faced with facility closure and serious financial difficulties. The state began to assume oversight and control over the university, but, at the same time, the federal government was investigating whether the state had provided adequate financial support to the university. The existence of this federal investigation was strategically used by Central State supporters to argue for the school's continued operation.

The disparate treatment of publicly supported HBCUs and HWIs by largely White legislatures is visible throughout the nineteenth and twentieth centuries. Eric Anderson and Alfred Moss (1999), for example, noted that the country had 27 Black colleges by 1915, including a number of state-funded agricultural and mechanical colleges. "[T]hese institutions . . . were dependent on usually hostile or indifferent White legislatures, which provided minimal funds and discouraged the development of serious college-level programs" (Anderson and Moss, 1999, p. 16). Marybeth Gasman (2004) has noted that the failure to fund Black colleges continued well into the twentieth century; "[the] chronic underfunding of Black colleges . . . created the need for the UNCF consortium," which led its first campaign in 1944. For CSU, the 1981 OCR report recounted a history of inadequate support for the school's development. First, the report expressed concern with the racial identifiability of student enrollment, faculty, and staff at Ohio institutions, coupled with a "[f]ailure to enhance Central State by providing fewer and lesser quality resources, and by duplicating Central State University's programs at nearby

State institutions" (Goggins, 1987, p. 113). In particular, the OCR concluded that the state's decision to build and enhance Wright State University in a location 11 miles away from CSU constituted racial discrimination in violation of Title VI (Fisher, 1998). The OCR found that the state maintained Central State "as an institution for Blacks," from which White students were dissuaded from attending through state action (Fisher, 1998).

Although the Office of Civil Rights complaint had been opened in 1981, action on it was halted until the mid-1990s because of both the Nixon and Reagan administration (Healy, 1996). Despite the failure of the federal government to take action regarding Central State, state officials responded to the growing financial woes of the 1980s with a complete lack of support. In September 1991, *Chronicle of Higher Education* reported that "Ohio Republican legislators are working to end subsidies to Central State University" (Jaschik, 1991). That year, the Ohio Senate voted to reduce, by half, a $15 million subsidy to Central State to help accommodate its high proportion of disadvantaged students. Although the subsidy was ultimately restored, state committees immediately began to review the need for any subsidy and state legislators prepared to argue for its reduction in the next funding cycle. Then-State Senator Eugene J. Watts, for example, called it "counterproductive" to maintain an HBCU, in light of the resources devoted to increasing minority participation at all institutions. Indeed, Ohio's funding allocation to Central State was cut by 15% from the 1990–91 school year to the 1992–93 year; Ohio State University, on the other hand, saw a decrease of only 9% during that same period (Jaschik, 1992).

By 1994, financial difficulties had led Central State trustees to consider the closure of Central State West, a branch campus serving a nontraditional student population of close to 500 with evening and weekend classes (Central State U. May Close a Campus, 1994). Former University President Arthur E. Thomas resigned in 1995 amid allegations that he misspent university funds on personal expenses and incurred large amounts of debt for the university (Healy, Jan. 1996). Three years later, an arbitration panel ordered Thomas to repay a portion of his severance package to compensate the university for misspent funds, but this award was later overturned at the trial court level.

According to Healy (1996), by 1996, Central State found itself with $6 million in debt, considering layoffs and program reductions. After a finding that the campus had mismanaged student aid funds, the US DOE threatened up to $4 million in fines. In December 1995, a $1.2 million advance from the state had been necessary to pay Central State's employees. The student population decreased from 3,300 in 1992 to 2,600 in 1996. Interim President Herman B. Smith Jr. immediately sought to reduce staff, improve fund-raising, and boost the state legislators' confidence in Central State. The faculty union responded with threats of lawsuits over the layoffs and salary reductions (Healy, May 1996). The state had also threatened to reduce its budget appropriation to the school unless it generated a surplus within six months (Healy, Jan. 1996).

By 1996, however, Ohio faced the enforcement of the OCR complaint that had been ignored for over a decade. Elaine Hairston, chancellor of the OBR, was a part of the state's vehement opposition to the assertion that their funding allocations to Central State had been anything other than legally appropriate and fiscally supportive. In a February 1996 letter to the editor of *Chronicle of Higher Education*, Hairston wrote, "Central State has had the highest level of state funding per student in the state for the past 15 years." This counterargument, however, ignored the bulk of the OCR's concerns, such as the development of Wright State University a mere 11 miles from Central State. Although Ohio maintained that its conduct toward Central State was consistent with federal civil rights requirements, by June 1996, the state had developed a plan to respond to the OCR's concerns (Healy, June 14, 1996). This proposal maintained Central State's mission of helping "underserved populations," imposed management procedures on the administration, emphasized science and education programs, developed graduate programs, and upgraded library holdings and campus facilities (Healy, June 14, 1996). Within a week of its submission, the OCR rejected the state proposal for failing to meet civil rights obligations (Healy, June 21, 1996). Specifically, the office was concerned that Ohio's plan did not set forth "clear, enforceable commitments" to the university, nor did it specify how the general intentions of the proposal would be funded. Raymond C. Pierce, then-deputy assistant secretary at the US DOE, stated that the OCR would be satisfied only when Ohio developed a plan to ensure that Central State would be "academically distinctive," receive long-term financial assistance, and develop new degree programs. Further, the federal government sought a public commitment by the state to keep Central State as a viable, publicly funded university.

As Ohio faced pressure from the federal government, Central State's difficulties worsened in July 1996 when the state fire marshal closed six dormitories, housing 1,300 students, because of health and safety concerns (Healy, July 1996). Further, the Ohio inspector general opened an investigation into Central State at the governor's request, joining the state auditor, state fire marshal, state architect, the OBR, and the Ohio attorney general in pending investigations on the university. Also that summer, interim President Smith was fired when Central State's debt did not decrease; his successor, Charles H. Showell Jr., resigned the position after only two months to return to his permanent job as dean of the College of Business Administration (Healy, Nov. 1996). The state investigated both Smith and Showell on the basis of allegations of mismanagement; both denied those allegations. By November, the university's debt had soared to $11 million, and enrollment had declined by more than 20%. Creditors threatened to terminate critical services, such as water service to the campus (Ohio Inspector General, 1997). Even as OCR demanded the state's commitment to keeping Central State viable and strengthening its programmatic offerings, some legislators advocated closing or reducing the programs of the school (Healy, Nov. 1996). In January 1997, then State Senator for the Ohio district in which Central State is located, Merle G. Kearns, proposed legislation that would allow the governor and state board of regents to take over financially struggling universities (Healy and Schmidt, 1997).

In February 1997, the Ohio inspector general had consolidated the numerous pending investigations into Central State originating in several state offices, and issued an interim report summarizing information from all of the Ohio entities that had been examining Central State. In its investigation, the state reviewed documents, subpoenaed witnesses affiliated with the university, conducted on-site visits, and coordinated with local prosecutors in case criminal activity was uncovered. The inspector general first concluded that the university's maintenance department was both underfunded and understaffed. Despite allegations by the OCR and in the media that the state provided Central State with inadequate funding, the inspector general clearly placed the blame for the maintenance-related budgetary problems on the shoulders of the university. For example, the inspector general's report focused on the university's denial of approximately 25% of the maintenance department's proposed budget in fiscal years 1993–96, as well as the refusal to hire additional requested personnel. Nevertheless, the report fails to discuss to what alternative uses those funds were put, although it does note that maintenance funds were directed to improvements at the president's residence. The inspector general "noted numerous instances in which the [maintenance] department's budget was used to purchase or renovate items located in the president's home [and] found requisitions for landscaping and renovations for the residence that would be, in our opinion, frivolous or unwarranted" (Ohio Inspector General, 1997, p. 12). The report, then, specifically chastised the university: "We question the wisdom of . . . spending for the president's home at, or about the same time, as safety and living conditions within the dormitories were being scrutinized by state inspectors. While thousands of dollars were spent sprucing up the interior and exterior of the president's residence between July 1995 and June 1996, students were living in squalid conditions" (Ohio Inspector General, 1997, p. 14).

Second, the inspector general noted that, between 1984 and 1994, Central State had complied over 100 separate citations from the state auditor, "ranging from poor record keeping to failure to follow standard accounting procedures" (Ohio Inspector General, 1997, p. 15). Many citations were imposed in successive years for the same violation. Again, the report concluded that sole responsibility lay with University officials.

> Each of the listed Auditor citations constitutes a failure to comply with state law and establishes acts of wrongdoing by university officials. We note our concern for the university's apparent inability or unwillingness to take appropriate corrective action when directed by state officials. Many of the specific failures cited were not expensive or labor intensive. The lack of funding or personnel is simply no excuse.
> (Ohio Inspector General, 1997, pp. 23–4)

Here, the inspector general was careful to rebut any potential argument that Central State's failure to comply with its record-keeping and recording obligations resulted from inadequate funding, likely because of the ongoing OCR investigation into state funding of the university.

Third, the report reviewed citations issued by the state fire marshal, and expressed particular concern over those code violations that were not corrected between a 1994 fire marshal report and a second report in 1996. Coupled with the repeated auditor citations, the inspector general concluded that the university had an "institutional history of rejecting [state] oversight efforts" (Ohio Inspector General, 1997, p. 24, 36). Similarly, the report later criticized the University Board of Trustees for failing to keep accurate and complete minutes of their meetings, leaving the state with little to review in terms of the board's role in providing financial oversight to university officials.

Fourth, the inspector general reviewed the correspondence between university officials and the OBR regarding the board's oversight of the university budget. According to the report, Central State reported a deficit of $3.6 million in December 1995.

Both the university and the board of regents termed the deficit, at this point, a "short-term cash-flow problem," and the board advanced 1996 funds to the university to enable it to meet its end of the year payroll obligations. Further, Chancellor Elaine H. Hairston directed Central State "to take immediate action to reduce expenses in an effort to bring them in line with revenues" (Ohio Inspector General, 1997, p. 38). In finding that the university failed to follow this directive, the inspector general cited a $12,000 expenditure for flowers and seeds, nearly half of which was spent on the president's residence, and a $32,711 expenditure on a security booth at the campus entrance that one former university official stated was excessive in light of the school's financial difficulties. The inspector general concluded that the university had made "an insincere effort" to solve its financial problems, noting that the investigation had revealed "many more examples" of this alleged insincerity (Ohio Inspector General, 1997, p. 40).

The report closed with the state's recommendations. First, Central State was to schedule regular preventative maintenance with the assistance of the state architect's office. Second, the board of trustees was instructed to prepare, approve, and distribute complete meeting minutes, as required under Ohio law. Third, the board was to "take appropriate measures" to comply with the recommendations and citations of the other state agencies that inspected the university, such as the state auditor and fire marshal. Additionally, the inspector general recommended that the university not approve any project to be performed by a private contractor without inspection by qualified maintenance personnel. The inspector general also included a recommendation to the state:

> There is a clear pattern of disregard for the findings and recommendations of oversight agencies that has transcended changes in administrations. It is our conclusion that "oversight" alone will not suffice. There must also be some level of interim "control" attached. Until Central State can demonstrate the ability to perform responsibility, its ability to expend state funds should be limited.
>
> (Ohio Inspector General, 1997, p. 44)

Indeed, this recommendation was followed by the Ohio General Assembly through subsequent provisions authorizing the state to appoint university officials and close the university if its directives were ignored.

Further, Central State's financial woes only worsened, with its skyrocketing debt at $20 million by April 1997 (Healy, Apr. 1997). OCR responded to state legislative efforts to close the university with threats to cease federal higher education funds to the state. Deputy Assistant Secretary Pierce expressed frustration with the fact that, despite its 1995 promise to develop a plan to improve Central State, the Ohio General Assembly was not considering a single proposal for such improvement by early 1997. Opposing state lawmakers charged that an increase in state funding would only be warranted if the university were financially sound and well managed. Then-Governor George Voinovich and supportive legislators were able to use the threat of OCR action to pressure Central State's opponents to preserve the university (Healy, Feb. 1998).

In 1997, the Ohio General Assembly passed two pieces of legislation in response to Central State's fiscal difficulties. Ohio Senate Bill 6 (1997) appropriated $10.3 million to reduce Central State's debt. That same bill required the development of rules authorizing the OBR to make a state university on "fiscal watch." Before a fiscal watch determination may be made, the state auditor would be required to notify several Ohio agencies and both legislative houses of a budget deficit. When a university is placed on fiscal watch, the governor may transfer university trustees' powers to a temporary conservator and suspend the president or chief officer of the university on watch. A university president can be reinstated by the conservator after a performance evaluation. Finally, the bill required the development of a training program for members of university boards of trustees and the OBR. Ohio Senate Bill 102 (1997) placed Central State "in a state of fiscal exigency . . . for fiscal years 1998 and 1999," and put in place spending prohibitions and reporting requirements during that period. The university board was ordered to phase out academic programs outside of the Colleges of Business, Education, or Art and Sciences, with a phase-out plan because of the board of regents by June 30, 1998. The Ohio Director of Budget and Management was directed to appoint a chief financial officer, and/or hire a financial supervisor to issue financial reports, until the director believes university officials are able to manage the school's fiscal affairs. The University Board of Trustees was required to consider faculty and staff reductions to balance the budget. The university treasurer had to post additional funds in bond before undertaking duties and the board of regents had to work with the university to enter into agreements with other Ohio universities for assistance in the development of the school's mission and fiscal solvency. Miami University was one such university, sending financial officers to Central State to develop fiscal management plans and procedures. Finally, the threat of closure loomed over the university. Senate Bill 102 required the director of budget and management to issue a remedial directive to Central State if it is substantially noncompliant with Ohio law. If the university failed to make "substantial, measurable progress after the remedial directive," the board of regents must close the university.

In July 1997, the state legislature authorized $28 million in funding over two years, as well as $6 million for facility repairs (Healy, July 1997). Although the funding package prompted the OCR to "suggest that a resolution . . . may be in sight," the package was tied to a number of benchmarks and conditions. Academic programs were to be reviewed and overhauled, the football and baseball teams were suspended, and layoffs were expected. Student enrollment was unusually low, as admissions officials had scaled back in anticipation of a potential closure. Although the average enrollment from 1965 through 1995 was 2,443, enrollment plummeted in 1996 to 1,910 students, then to a low of 973 in 1997 (Ohio Board of Regents, 1997). Although enrollment has steadily risen since 1997, it has remained below 2,000 students in subsequent years; current 2005 enrollment was 1,623 (Ohio Board of Regents, 2005).

Also during the summer of 1997, Central State alumnus John W. Garland left his position as associate vice-provost at the University of Virginia to take over as permanent president of Central State (Strosnider and Healy, 1997). Garland took over at the start of the 1997–98 school year, and has remained in this position to date. In February 1998, OCR closed its investigation of Ohio's treatment of Central State based on the continued financing and restructuring plan (Healy, Feb. 1998). Although Norma V. Cantu, assistant secretary for Civil Rights, concluded that Ohio might not yet be in compliance with federal civil rights requirement, OCR believed that compliance would be reached with full implementation of the funding and restructuring plan. Deputy

Assistant Secretary Pierce explicitly stated, "There is no determination that the violation has been corrected at the time of the closure" (Fisher, 1998).

In March 1998, the OBR announced that the Office of Budget and Management "officially released the financial management of Central State University to the Board of Trustees and University administration" (Ohio Board of Regents, 1998). Nevertheless, the years of financial difficulty and wavering state support had taken its toll. Between 1996 and 1998, enrollment fell by close to 60%, the 121-member faculty was cut to 82 members, and only 30 of 50 major offerings remained (Ohio Board of Regents, 1998). By November 1998, questions over state funding adequacy resurfaced, as the state 1998–99 budget did not include additional funds to Central State despite nationwide higher education funding increases (Schmidt, 1998). In March 1999, the state's investigation into the acts of former Central State presidents Thomas and Smith concluded with a finding that both had "committed acts of wrongdoing and omission resulting in mismanagement of the university's academic, fiscal, and physical plant operations" (Ohio Inspector General, 1999). The report concluded with eight administrative recommendations for the university. Between 1998 and 2002, Central State's administration, a team put in place by the state of Ohio after the 1997 takeover, balanced the budget, repaired facilities, and earned clean audits for three years (Schmidt, 2002). University enrollment reached 1,400 in 2002, a close to 30% increase from 1997. In April 2002, Governor Bob Taft announced that the state's financial oversight was no longer warranted.

CENTRAL STATE'S CONTINUING VIABILITY

Nevertheless, the years of state oversight and fiscal difficulties continued to impact the campus. It was not until the 2005–06 school year that the school had its own football team again, after a nine-year hiatus (Garland, 2005). Indeed, the suspension of the school's football team took from Central State an important part of its identity. Patrick Miller (1995, p. 117) noted the importance of the sport's introduction to Black colleges in the late nineteenth and early twentieth centuries: "A game like football offered memorable examples of competence and vitality just as it represented one of the principal measures of institutional prestige, both within the African American community and beyond." In 1893, one student commented that "to excel in athletics . . . [would] raise the honor" of a Black college (Miller, 1995, p. 118). Students at Central State in 2005 felt no different. Indeed, Wilberforce had a long history of athletic promotion. According to Miller (1995, p. 118), "though the inauguration of off-campus athletics was often in reality quite a modest affair, through memory and nostalgia it became a prominent part of the early histories of schools from Wilberforce in Ohio to Talladega in Alabama."

Enrollment numbers dropped again, with an 11% decrease between 2004–05 and 2005–06 (Garland, 2005), despite the school's goal of returning to an enrollment of over 3,000 students by 2010 (Cox News Service, 2005). According to Garland (2005), state funding also declined.

In June, [Ohio] Governor [Bob] Taft signed into law a budget bill that reduced CSU's funding by $1.6 million over the next two years. Since 2002, when

Governor Taft removed Central State from fiscal watch, our state funding has decreased by almost 14 percent, from $19.3 million to $16.5 million in FY2006.

In response to the declining state funding, President Garland (2005) pointed to the possibility of reopening the OCR investigation in a public letter:

> In 1998, the State of Ohio pledged to enhance Central State to make it as attractive as and comparable with other public universities in Ohio. This pledge formed the basis of an agreement that settled a decades-long [OCR] investigation . . . into the state's non-compliance with Title VI of the Civil Rights Act of 1964. . . .
>
> After the 1998 agreement, CSU saw a significant increase in state funding for a couple of years, but most of that went to take care of deferred maintenance, itself the result of years of reduced state capital support. The reductions in funding since 2002 have limited our ability to do things such as start new programs, add new faculty, and improve our facilities. All play an important part in attracting and retaining students. We should be budgeted primarily for excellence and growth, rather than budgeting to balance the budget.

In his public letter, then, CSU President Garland made the immediate link to the 1998 closure of the OCR investigation. In comments to the off-campus media, Garland also drew an explicit link between declining state funding and declining student enrollment. A local newspaper reported, "Garland said some potential students are being priced out of a college education at CSU because of tuition increases brought on by state funding cuts" (Cox News Service, 2005). Indeed, Garland informed the public that, in December 2004 and at the direction of the CSU Board of Trustees, he formally requested that OCR reopen their investigation into Ohio's treatment of the school (Garland, 2005). OCR responded with a denial of the request to reopen, but with a promise to continue monitoring the state's compliance. Further, Garland (2005) hinted at continued correspondence with the OCR and even possible future litigation: "Citizens frustrated by state inaction in Mississippi and Texas filed and won federal lawsuits aimed at getting those states to comply with Title VI. We have tried a different tack, pleading our case before state leaders and policymakers. The results, so far, have been disappointing, not to mention frustrating."

This tactic of relying on the potential influence of the federal government over state appropriations appears to be one that comports with Central State's experience in the 1990s. Indeed, when the OCR mandate was not enforced in the late 1980s and early 1990s, the state of Ohio threatened to reduce or end the university's state subsidies. After the OCR resumed attention to Central State in 1996, and particularly after its threat to discontinue federal higher education funds for Ohio if compliance was not met, the state put together an aid package to help the struggling university recover from its financial difficulties. Finally, after the OCR investigation was closed in 1998, the state quickly resumed funding Central State at the lower level of earlier years.

Indeed, Central State did see a relative increase in state funding after the Ohio–OCR agreement was reached, but saw a decline in that funding within two years of the OCR investigation's closure. Figure 9.1 demonstrates two interesting points. First, as the OBR has long maintained, Central State does receive the highest amount of per-student funding in the state; nevertheless, this figure has never been

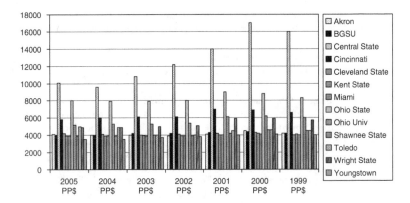

Figure 9.1   1999–2005 per-pupil state funding for operating expenses.

the focus on the OCR investigation nor the basis for its conclusion that Ohio's treatment of Central State violates Title VI. From a legal standpoint, then, this counterargument is effectively irrelevant. Second, even per-student spending, which is the OBR's chosen measure of fiscal equity, sharply declines within two years of the OCR investigation's closure. This funding pattern at least raises the question of whether the state's support of Central State was primarily, or even solely, a response to that federal investigation. What makes this potentiality fascinating is that many have charged that Title VI and higher education desegregation jurisprudence could lead to the closure of HBCUs. For Central State, it appears that Title VI may have kept the school open.

Nevertheless, Central State's position of reliance on OCR as an advocate is precarious at best. First, the enforcement strategies of OCR, and the attention that the agency has been willing to give Central State, vary with each administration. The original OCR finding that Ohio had failed to comply with Title VI was made in 1981, but OCR officials within Reagan and George H. W. Bush administrations did not act on this finding. It was not until several years into the Clinton administration that OCR action was finally taken. If Central State's continued existence is heavily dependent on federal oversight, this may be dangerous for the university. Second, the plans developed in the 1996–98 period, and supported by the OCR, to resolve the school's financial difficulties and meet Title VI compliance under the OCR investigation involved, essentially, downsizing (Fisher, 1998). In 1998, state officials defended faculty layoffs and academic department mergers by arguing that they were necessary to secure CSU's financial stability, but that development would be forthcoming. Then-State Senator Jeff Johnson from Cleveland, a leader on legislative efforts to save CSU from closure, asserted that downsizing "had to occur to get the business and finances in order and to get the school funded for the next two years. Sometimes you have to take two steps back to take three or four steps forward" (Fisher, 1998). Both NAACP and CSU faculty union representatives expressed concern about the implicit federal approval of this approach with the 1998 closure of the OCR investigation. A CSU official spoke to the press of similar concerns, although under conditions of anonymity. The original violations that led to the OCR investigation, however, were based on the argument that

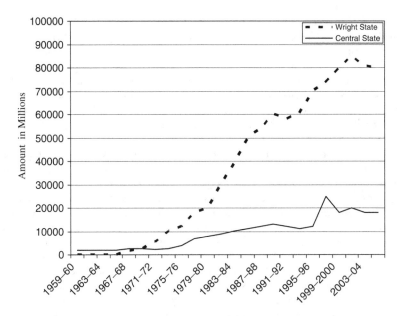

Figure 9.2   State of funding to Central State and Wright State, 1959–2005.

state funding had halted CSU's opportunity to grow as a university, specifically in favor of nearby Wright State University. Figure 9.2 compares state funding levels for the two universities from 1959 through the present.

The difference in state funding increases is outstanding, particularly when one considers that Wright State only began receiving funding in 1966 and opened a mere 11 miles from Central State. The state of Ohio's differential treatment of Wright State, an HWI, and Central State, an HBCU, was a primary indication to the OCR of a failure to comply with Title VI in 1981. Nevertheless, nothing has been done by the state, or demanded by the OCR, to counteract this difference. Again, if Central State's continuing existence rests on an active inquiry by the OCR, the school is not in an enviable position. The OCR's attention to Central State has wavered greatly over the past two decades, and, even in periods of its most ardent advocacy, the office has demanded little to combat the apparent state policy of limiting or reversing Central State's growth.

CONCLUSION AND RECOMMENDATIONS

Much is unique about Ohio's only publicly supported HBCU, CSU. Unlike the Southern public HBCUs, Central State did not develop within a system of de jure segregation. Indeed, Wilberforce, the predecessor to Central State, first received public funds in the same year that Ohio dismantled racial segregation in schools. Despite Central State's development under a de facto system of segregation, there are similarities with those HBCUs that were founded under de jure segregation, particularly evident in an examination of the contemporary treatment of publicly supported HBCUs.

For example, one of the four constitutionally problematic elements of Mississippi's formerly segregated higher education system, according to the *Fordice* Court, was that of the geographical proximity between HBCUs and HWIs. In Ohio, just over a decade after *Brown* and within two years of the passage of Title VI, the state opened and heavily funded an HWI only 11 miles from Central State. It is clear, then, that despite the differences between states that did not require racial segregation and those that did so mandate, there are similarities in the fiscal treatment of HBCUs. Further comparative research into state funding for HBCUs in formerly segregated states, as opposed to states with de facto segregation only, would be interesting.

In addition, the type of unequal treatment observed in the Central State example conflicts with federal civil rights legislation and higher education desegregation jurisprudence. Although commentators have questioned whether these laws would lead to the closure of HBCUs, Central State officials have been able to rely on an OCR investigation into the state's treatment of their school in arguing for the school's continued existence despite a financial crisis. Given that state financial support has declined since the closure of the federal investigation, however, it is not clear whether such an investigation must be ongoing for the state of Ohio to meet its legal obligations to CSU. Further, the OCR-supported plan ultimately did little to counteract decades of underfunding by the state. Indeed, the state of Ohio kept Central State's funding allocation significantly lower than nearly all other state-supported universities, which resulted in the stunting or even the reversal of Central State's growth in favor of the development of HWIs. Even during the period of heightened OCR involvement, little was done to foster Central State; the OCR was satisfied with a one-time influx of aid to enable the school's continued operation only.

Therefore, there is some evidence that publicly funded HBCUs should cultivate their relationship with and support from federal agencies, such as OCR, which appear to be more sympathetic to the continued existence of HBCUs. On the other hand, it is also apparent that HBCUs cannot rely on the OCR to ensure state funding for institutional growth. That said, for Central State, the predictions of rampant HBCU closures following legal developments on desegregation and civil rights were not well founded. Instead, a federal civil rights investigation assisted the university in maintaining its place in Ohio's higher education system.

NOTE

1. The literature on the early years of Central State University is sparse. This section relies heavily on the work of Lathardus Goggins, and, where a statement is uncited, it should be presumed that Goggins (1987) is the original source.

REFERENCES

Adams v. Richardson, 480 F.2d 1159 (D.C. Cir. 1973).
Anderson, E., and A. Moss Jr. 1999. *Dangerous donations: Northern philanthropy and Southern Black education, 1902–1930.* Columbia, MO: University of Missouri Press.
Author unknown. 1994. Central State U. May Close a Campus. *Chronicle of Higher Education,* February 16. Retrieved December 2005, from http://www.chronicle.com.

Bowles, F., and F. DeCosta. 1971. *Between two worlds: A profile of Negro higher education.* New York: McGraw-Hill.

Brown v. Board of Education (Brown I), 347 U.S. 483 (1954).

Brown v. Board of Education (Brown II), 349 U.S. 294, 301 (1955).

Center for the Study of Education Policy, Illinois State University, University Budget Data. Retrieved December 2005, from http://www.coe.ilstu.edu/grapevine.

Cox News Service. 2005, October 10. CSU enrollment off about 11 percent: Expected boost from football's return didn't happen. *Middletown Journal.* Retrieved December 2005, from http://www.middletownjournal.com.

Cummings v. Richmond County Board of Education, 175 U.S. 528 (1899).

Fisher, M. 1998. Feds close Title VI investigation in Ohio. *Black Issues in Higher Education.* Retrieved December 2005, from http://www.findarticles.com.

Gaines v. Canada, Registrar of the University of Missouri, 305 U.S. 337 (1938).

Garland, J. W. 2005, October 17. Letter. *Gold Torch: CSU Student Newspaper.* Retrieved December 2005, from http://www.goldtorchnews.com.

Gasman, M. 2004. Rhetoric v. reality: The fundraising messages of the United Negro College Fund in the immediate aftermath of the Brown decision. *History of Education Quarterly* 44(1): 70–94.

Goggins, L. 1987. *Central State University: The first one hundred years, 1887–1987.* Wilberforce, OH: Central State University.

Healy, P. 1996. President of Central State U. considers cuts and layoffs. *Chronicle of Higher Education,* January 19. Retrieved December 2005, from http://www.chronicle.com.

———. 1996. Public historically Black colleges face myriad of problems. *Chronicle of Higher Education,* May 17. Retrieved December 2005, from http://www.chronicle.com.

———. 1996. Civil Rights Office reviews Ohio plan to improve Central State. *Chronicle of Higher Education,* June 14. Retrieved December 2005, from http://www.chronicle.com.

———. 1996. Ohio told to provide details on plan to improve Central State. *Chronicle of Higher Education,* June 21. Retrieved December 2005, from http://www.chronicle.com.

———. 1996. Fire marshal closes Central State U. dormitories. *Chronicle of Higher Education,* July 19. Retrieved December 2005, from http://www.chronicle.com.

———. 1996. Another acting President steps down at Central State U. *Chronicle of Higher Education,* November 22. Retrieved December 2005, from http://www.chronicle.com.

———. 1997. U.S. tells Ohio it must improve Central State U. or lose funds. *Chronicle of Higher Education,* April 4. Retrieved December 2005, from http://www.chronicle.com.

———. 1997. Ohio lawmakers keep Central State alive, but demand overhaul. *Chronicle of Higher Education,* July 3. Retrieved December 2005, from http://www.chronicle.com.

———. 1998. Education department ends investigation of Ohio's treatment of Central State U. *Chronicle of Higher Education,* February 27. Retrieved December 2005, from http://www.chronicle.com.

Healy, P., and P. Schmidt. 1997. Public colleges expect tough rivals in annual fights for funds. *Chronicle of Higher Education,* January 10. Retrieved December 2005, from http://www.chronicle.com.

Jaschik, S. 1991. Future financing at stake: Legal and philosophical positions of public Black colleges at a watershed as high court takes up desegregation case. *Chronicle of Higher Education,* September 25. Retrieved December 2005, from http://www.chronicle.com.

———. 1992. 1% decline in state support for colleges thought to be first 2-Year drop ever. *Chronicle of Higher Education*, October 21. Retrieved December 2005, from http://www.chronicle.com.

Miller, P. 1995. To "bring the race along rapidly": Sport, student culture, and educational mission at historically Black colleges during the interwar years. *History of Education Quarterly* 35(2): 111–34.

Ohio Board of Regents. 1997. *Basic data series historical data, table 2: Student enrollment, fall headcount, Ohio public institutions.* Retrieved December 2005, from http://www. regents.state.oh.us.

———. 1998. *The regents' report*, 1(1). Retrieved December 2005, from http://www.regents. state.oh.us.

———. 2005. *Headcounts at public colleges and universities, 2004–2005.* Retrieved December 2005, from http://www.regents.state.oh.us.

Ohio Inspector General. 1997. *Interim report of investigation, Case No. 96–114-IG.* Retrieved December 2005, from http://www.watchdog.oh.gov.

———. 1999. *Annual report: Selected summaries of investigations in 1999, Case No. 1996114, Central State University.* Retrieved December 2005, from http://www.watchdog.oh.gov.

Ohio Senate Bill 102. 1997. Retrieved December 2005, from http://www.legislature.state. oh.us.

Ohio Senate Bill 6. eff. March 21, 1997, and June 20, 1997. Retrieved December 2005, from http://www.legislature.state.oh.us

Plessy v. Ferguson, 163 U.S. 537 (1896).

Samuels, A. 2004. *Is separate unequal?* Lawrence, KS: University Press of Kansas.

Schmidt, P. 1998. State spending on higher education rises 6.7% in 1998–99 to a total of $52.8 billion. *Chronicle of Higher Education*, November 28. Retrieved December 2005, from http://www.chronicle.com.

———. 2002. Ohio ends oversight of Central State U. *Chronicle of Higher Education*, May 3. Retrieved December 2005, from http://www.chronicle.com.

Strosnider, K., and P. Healy. 1997. Troubled Central State U. names alumnus as president. *Chronicle of Higher Education*, July 25. Retrieved December 2005, from http://www. chronicle.com.

Taylor, E., and S. Olswang. 1999. Peril or promise: The effect of desegregation litigation on historically Black colleges. *Western Journal of Black Studies* 23(2): 73–82.

Tucker, S. K. 2002. The early years of the United Negro College Fund, 1943–1960. *Journal of African American History* 87(4): 416–32.

United States v. Fordice, 505 U.S. 717 (1992).

Williams, J. 2004. *I'll find a way or make one: A tribute to historically Black colleges and universities.* New York: Amistad/HarperCollins.

Women's Equity Action League v. CAVA205, 906 F.2d 742 (D.C. Cir. 1990).

CHAPTER 10

THE RETENTION
PLANNING PROCESS
AT TEXAS SOUTHERN
UNIVERSITY:
A CASE STUDY

JACQUELINE FLEMING, ALBERT TEZENO,
AND SYLVIA ZAMORA

Retention is a major issue at Texas Southern University (TSU), as it is at most universities. As a historically Black public institution that has an open admissions policy, the effort to increase retention rates occupies considerable attention. This chapter is based on an institutional study of a planning process designed to improve the retention of freshmen at TSU.

Researchers report average dropout rates from college for African Americans ranging between 63.3% and 75%, but reaching as high as 90% at large state universities and open admissions institutions (Porter 1990). Among White students the retention rates are no better than 50%, but they are still better than that of minorities (Tinto, 1987; Richardson and Bender, 1987; National Center for Educational Statistics, 2007). According to Tinto (1975; 1987), students do not drop out of college primarily for academic reasons. They drop out for the same reasons that people commit suicide, because they are not integrated into the fabric of society. Thus, the many factors that might increase the disconnectedness from the academic and social fabric of college life hold the key to retention. This approach makes intuitive sense for African American students. Their racial difference makes for a racial divide. The separatism, alienation, and discrimination that often ensue would constitute disconnecting factors. There is indeed research that supports the integration theory of dropping out among African American students (Fleming, 1984; Morning and Fleming, 1994; Pascarella and Terenzini, 1977; Tinto, 1987). Does, however, the integration theory of retention hold for African American students in predominantly Black institutions? According to the United Negro College Fund (UNCF), retention is as much an issue for Black colleges as for historically White institutions (HWIs) (UNCF, 1998). While retention theory contends that academic failure is not the essential cause of attrition, academic underpreparedness may be more of a retention issue for African American students, especially in Black colleges. Black students appear to experience more academic problems than White students, even when they have similar abilities (Sherman et al., 1994; Mow and Nettles, 1990). In general, the academic preparation of students attending historically Black colleges and universities (HBCUs) falls below that of students in other institutions (DeSousa, 2001; Galloway and Swail, 1999). While the graduation rates at Black colleges have been assailed for being lower than at other institutions, Astin, Tsui, and Avalos (1996) found that among students of similar ability levels, Black students at Black institutions were actually more likely to graduate. These authors suggest that institutional size may be the primary reason, but Fleming (2002a) found that Black institutions offer freedom from nonacademic distractions, such as racism. Nonetheless, the problem of academic underpreparedness exerts a major impact on the retention rates of Black institutions, such as TSU.

President Priscilla Slade examined TSU's retention rates in comparison with other public institutions in Texas. Figure 10.1 shows that the 1997 one-year retention rates at TSU were lower than at other public institutions in Texas. This is largely because so many incoming students have not had a college preparatory curriculum and have not passed all parts of a Texas state–mandated college entrance test. In 1997, TSU retained only 41.5% of freshmen who had regular high school diplomas and had not passed all three parts of the Texas Academic Skills Program (TASP) test in math, reading, and writing. TSU ranked last in comparison with selected public institutions in Texas, and below the state average. The problem is that in any given year, from 50% to 80% of incoming students have not passed all parts of a Texas college entrance test. At TSU, such students are referred to as test responsible.

Figure 10.2 shows that when college-ready students are considered, TSU's one-year persistence rates were above the state average and ranked fourth among selected public institutions. TSU retained 82.8% of college ready students with advanced (i.e., college)

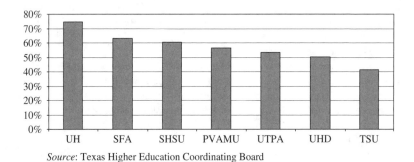

Source: Texas Higher Education Coordinating Board

Figure 10.1   Retention of first-time full-time freshmen with *regular* diplomas who are not college ready at Texas Southern University compared with public institutions in Texas, 1997.

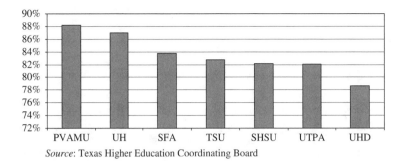

Source: Texas Higher Education Coordinating Board

Figure 10.2   Retention of first-time full-time freshmen with *advanced* diplomas who are college ready at Texas Southern University compared with public institutions in Texas, 1997.

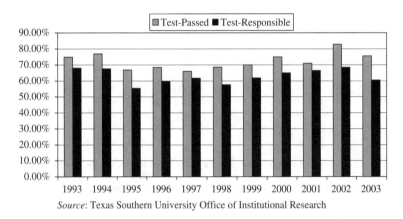

Source: Texas Southern University Office of Institutional Research

Figure 10.3   One-year retention rates for test-passed and test-responsible first-time freshmen entering in 1993 through 2003.

diplomas from freshman to sophomore year. Thus, TSU's overall weak performance on key outcome measures is largely a function of the level of preparation of the students who come to the university.

President Slade undertook a 13-year analysis of TSU's retention rates. Figure 10.3 shows that since 1997, the one-year retention rates have improved steadily. Last year, the rates took a dip that may be due to a 12% increase in tuition. Note that the retention rates for test passed students are good—up to 82%. The rates for test responsible freshmen average 63% and range up to 68.5%. The current overall retention rate is 62.9%, which compares favorably with other four-year open-admission institutions nationwide (Fleming et al., 2005). While the trend toward increasing retention rates is a good sign, it is all the more remarkable when the qualifications of incoming students are considered.

Figure 10.4 shows the 13-year trends in Texas Academic Skills Program (TASP)/Texas Higher Education Assessment (THEA) test scores and indicates that

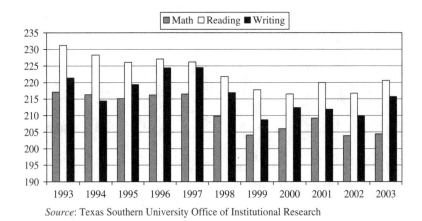

Source: Texas Southern University Office of Institutional Research

Figure 10.4   Trends in TASP math, reading and writing scores for first-time freshmen entering in 1993 through 2003.

since 1997 the scores have declined steadily, especially in math. Institutional evidence indicates that the math score is the best single predictor of one-year retention (Fleming, 2004). So, despite declining test score qualifications of entering students, retention rates have increased. Despite increases in retention, TSU seeks to improve its retention record.

RETENTION PROGRAMS AT TEXAS SOUTHERN UNIVERSITY:

THE GENERAL UNIVERSITY ACADEMIC CENTER (GUAC)

The increases in retention rates despite declining test scores have not occurred by accident. TSU devotes considerable energy to retention planning. For example, each of its eight colleges has a detailed retention plan, and each college has its own academic support program or plans to institute one. The Office of Student Assistance offers an array of academic support programs to all TSU students. Furthermore, the General University Academic Center (GUAC) offers its own array of academic support programs to freshmen in general and to test responsible students in particular.

GUAC academic support programs are guided by the academic literature and the best practices indicated therein. GUAC offers the 15 academic support programs. The Bridge Builders Peer Tutoring program is certified by the National Tutoring Association at University of Missouri at Kansas City (UMKC). This program is in the process of being recertified by the College Learning and Reading Association, also at UMKC, in accordance with National Association of Developmental Education guidelines (NADE). The Supplemental Instruction Program is also certified by UMKC. These two programs and two additional ones, GUAC 101 Adjunct Instructional Programs and the Freshman Year Transition Course (FYTC), are in the process of NADE certification.

Academic advisement takes place during registration, by appointment, online, by e-mail, and through pop-up notices. In addition, academic advisors go to the student center each week to work with students where they are most likely to gather. The Bridge Builder's Peer Tutoring program employs trained honors students to tutor students in a variety of subjects. The Supplemental Instruction program enables trained Masters students (who are shared by the tutoring staff) to conduct review sessions after class in gatekeeper courses with high failure rates, such as college algebra and freshman English.

The Learning Specialist's function is to teach students how to study and how to improve upon their study and academic management skills, since many students come to college without the experience of having studied in high school. The Intensive Academic Support program for students on probation provides a last chance for students on the verge of dropping out for academic reasons to stay in school. The program requires mandatory visits to tutoring, the Learning Specialist, academic advisors, and Study Skill Enhancement seminars.

The annual Major Fair targets students with undecided majors and provides access to advisors, fellow students, and information from a wide array of major subjects. Vocational interest testing is available on the spot. Furthermore, students attending the Major Fair are monitored and sent relevant information on majors and programs to help the undecided through the year. Mentoring.com pairs students in sections of the FYTC with on- and off-campus mentors who communicate by e-mail. A peer-mentoring program will begin this fall where students e-mail a cohort of incoming students to guide them through critical aspects of the first-year experience. The Community Service Assistance Program provides career exposure through a six-week volunteer internship experience through sections of the FYTC. Interns work at selected business sites throughout Houston to gain an appreciation of what the envisioned career is like in reality.

GUAC has instituted quiet Study Hall at specified hours during the week. It is open to all students but required of band members. During Study Hall, students also have access to tutors and to the Learning Specialist. The Summer Academy provides incoming test responsible students with an eight-week boot-camp like opportunity to prepare for college level work. The Summer Academy specializes in a number of interactive teaching methods designed to keep students engaged in the learning process, such as the Socratic Method of Q&A, group competitions, and the Whimbey and Lochhead (2000) method of Talk Aloud Pair Problem Solving (TAPPS). Summer Academy instructors are trained in these methods. New Student Orientation is the first formal introduction to TSU where students hear motivational keynote speakers, registration procedures, and attend a series of conference breakout sessions tailored to the various special needs of students. GUAC 101 Adjunct Instructional programs in reading, math, and writing are required tutorials for students with near passing test scores to help them succeed in college level classes. Sociology 211, TSU's FYTC, is a one-hour course that provides instruction in study skills and paths to academic success.

HOW EFFECTIVE ARE GUAC ACADEMIC SUPPORT PROGRAMS?

GUAC carefully monitors student participation, satisfaction, grade point average (GPA), and year-to-year retention for students enrolled in its programs. Some GUAC programs are full to capacity, such as the FYTC and GUAC 101 Adjunct Instructional programs in math, reading, and writing. But other programs are undersubscribed and

Table 10.1   Selected GUAC program statistics, 2003–2004

Program	N	Percentage in good standing	1-year retention (%)	High attendees 1-year retention (%)
GUAC 101 Math	374	50.0	64.9	72.0
GUAC 101 Reading	107	20.6	69.2	40.0
				N = 5
GUAC 101 Writing	1,927	52.6	57.8	69.2
FYTC-Soc 211	1,391	42.9	60.9	66.3
Band Study Hall	83	43.4	77.1	93.1

could service more students. Even in full programs, enrolled students may not participate regularly.

An examination of the average semester GPA of students enrolled in GUAC retention programs shows that it hovers around 2.0 or less for programs with students having special needs. The average one-year retention rate is about average—62%. In an effort to improve the effectiveness of these programs, the statistics were examined in greater depth.

When programs were analyzed by attendance or participation, performance, and retention was usually better for high attending and participating students as might be expected (Fleming et al, 2004). However, it became clear that 30% or fewer students participated regularly in any of the programs. In GUAC 101 courses, for example, only 10% participated regularly, with regular participants having better pass rates, first semester GPA, and one-year retention rates. Table 10.1 shows one-year retention rates for students in selected programs in comparison with retention rates for high attendees. Retention rates for high attendees were higher than average, and were as high as 93.1% for Band Study Hall members. The only instance in which higher rates were not found was for GUAC 101 Reading, where there were only five students in the high attending group. In subsequent analysis of this program where there were more students in the high attending group, their performance was higher than average.

We do not necessarily believe that high participation in GUAC programs is the sole reason for better student performance. The willingness to participate in educational programs may be an indication of higher student motivation (Thomas and Higbee, 2000; Friedman, Rodriguez, and McComb, 2001). Motivation to attain an education may be the biggest factor in performance. Nonetheless, it does seem reasonable that higher participation would serve the interests of retention. GUAC has searched for ways to improve program participation, including conducting advertising campaigns that emphasize the performance results just described.

A Retention Plan that Cannot Fail: The Revamped Plan

With the results from this monitoring effort in mind, President Slade challenged GUAC to design a retention plan for freshman that could not fail. It had to be obvious that the plan would be effective, especially with test responsible freshmen. The result was a densely packed 11-page plan grounded in the retention literature as well as evidence of what worked with TSU students. The plan had three basic components: (1) Mandatory Learning Assistance Programs—that is, students had to be required to participate in the academic support programs that had proven to be effective; (2) The communication of consistently high expectations by all members of the academic community—in contrast to the criticism and belittling that students so frequently report; and (3) The redesign of freshman instruction to create highly interactive classrooms. This last component was based on widespread evidence that TSU students today are often unable to learn from lecture alone without a method of hands-on engagement in the learning process. The plan was consistent with the universal formula for academic success: set high expectations with support for reaching those expectations.

As far as interactive classrooms were concerned, GUAC has experimented with a number of methods of achieving them because it is very obvious to any classroom observer that students do not learn well from traditional teaching methods. The Fast Track Program and the Summer Academy Program have been experimental laboratories for these methods. The Socratic Method is used as a basic technique of classroom management, so that questions are asked systematically of students who are called on multiple times by name during the class session. Instructors were trained in the art of constructive feedback because so many students complained that instructors criticized and belittled them for what they did not know. GUAC borrowed the TAPPS method from Whimbey and Lochhead (2000), where a student comes to the board to solve a problem and is prompted by the instructor to talk through its solution. This allows the student's thinking process to become visible and helps other students learn how to approach problems. GUAC used a number of group work scenarios and game formats to permit team cooperation. A number of multicultural instructional materials have been developed in both reading and math. Team competitions were recently instituted to help students translate their skills into a slightly more practical arena. For example, in math students compete on logic problems, which is an extension of mathematical skill. At this point, there is programmatic evidence that interactive classrooms and multicultural instructional materials facilitate performance advantages (Fleming, 2004, 2005a, 2005b; Fleming et al., 2004).

Although GUAC has developed a retention plan for freshmen that in theory should be effective, there were two problems. First, student participation in the academic support programs that would help them could not be mandated. Second, instructors could not be required to use these or other methods of more effectively reaching their students. GUAC could only set an example and discuss promising methods in an open forum. Because of the lack of power to mandate these changes, the retention plan was in limbo and in danger of losing momentum. It was at this point that a TSU team was invited to attend the USA Funds Symposium for Minority-Serving Institutions (MSI).

USA FUNDS SYMPOSIUM FOR MINORITY-SERVING INSTITUTIONS

A team of four individuals from TSU was invited to attend the USA Funds Symposium. USA Funds is a financial institution concerned with the default rate on student loans. In an effort to reduce default, USA Funds offers this symposium to assist institutions in developing and refining effective retention plans. The effect of the symposium was to jumpstart a retention planning process at TSU that was about to stall, despite previous successes. Conference participants were assembled into groups of about four institutions. At the opening workshop the Affinity Diagram Process was introduced, and then completed in two additional group workshops. This process was complemented by sessions on other retention plans and outcomes, goal-setting strategies, how to evaluate retention plans, and completing the USA Funds retention plan template.

THE AFFINITY DIAGRAM PROCESS DEFINED

The Affinity Diagram Process is a structured brainstorming tool that allows teams to identify issues, particularly when the topic is large and complex. Developed by a Japanese social scientist, Kawakita Jiro, as part of the Total Quality Management movement, the affinity diagram is designed to focus a team on *problems and issues, not solutions* (Tague, 2004). Ideas are generated, clarified, grouped together, and summarized. The resulting diagram then allows teams to generate prioritized solutions for each grouping. There are 15 steps in the process. The most notable of these are idea generation, grouping, generating titles for the groups, prioritizing the ideas, and summarizing the outcome.[1]

For *idea generation,* Post-its were distributed to each participant so that the total number of ideas generated is about 30. For example, ten people in the group, three Post-its each. Following the open discussion, each participant writes down an idea that addresses the theme. One idea per Post-it is written as a complete sentence, focusing on problems and issues, not solutions. As Post-its are completed, participants stick them on the left side of the flip chart.

For *grouping,* participants are instructed to group the ideas together. Participants should not necessarily overthink this process but look for groupings at a "gut level" because these items seem to belong together. The process is done in silence. Everyone stands and starts moving the Post-its around. Participants may move things back and forth. It also is permissible to have an idea by itself, a "lone wolf." When participants are satisfied with the groupings, they sit down.

To *generate titles for the groupings,* moderators read the cards in each group to the participants who, as a group, generate a title for the group—again using a full sentence stating a problem, not a solution. Titles are not necessary for "lone wolves." Moderators write the titles on Post-its using a red marker, and then stack the original Post-its under the title Post-it so only the title is showing.

To establish *priority,* each member gets a red dot (3 points), a blue dot (2 points), and a green dot (1 point). They vote at the title level for the grouping they think is most important as it relates to the theme, red (3) being the highest. They can use their dots any way they want, as long as everyone votes at once—no holding back to sway the outcome. The points are totaled. Groupings with the highest number of points are outlined in red, followed by blue and green outlines.

To *summarize the outcome,* the group comes up with a statement that summarizes the diagram and writes the summary in the upper right hand corner of the flip chart.

SELECTED GROUP SUMMARY STATEMENTS

Table 10.2 presents affinity diagram summary statements from the 16 groups represented at the symposium. For example, group 4 had nine groupings, four of which

Table 10.2   Affinity Diagram summary statements concerning retention for minority-serving institutions attending USA Funds Symposium

Group 1: The goal of student retention must permeate every aspect of institutional life. That results in graduation.

Group 2: It takes the students' community and early intervention to maintain and increase student retention.

Group 3: Retention is an institutional responsibility.

Group 4: Student retention is associated with the institution's academic resources combined with the individual student's preparation and background.

Group 5: Who are you? What do you offer? How does it benefit the "customer"? How would the customer qualify?

Group 6: Students are not adequately prepared to deal with the college transition, as well as institutions are not sufficiently prepared to meet their advising, financial, social, and personal needs.

Group 7: Higher education is a public good and a partial public responsibility.

Group 8: Retention requires strong institutional support systems to overcome a lack of academic preparation and unmet financial need.

Group 9: Same as group 8.

Group 10: Lack of institutional resources, academic preparation, and competing student priorities are the primary retention issues.

Group 11: As it relates to retention, institutional commitment and inadequacy of resources are important factors. However, students' competing priorities combined with a lack of academic preparedness greatly influences lack of student engagement.

Group 12: Retention must be a top institutional priority that requires continuous collaboration among academic, student, and business affairs.

Group 13: Lack of student preparation is central to retention problems, and there are institutional, financial, and personal impediments to overcoming a lack of preparation.

Group 14: American Indian student success is achievable when students are academically prepared, have an educational and career goal, and when the institution has the resources to provide the services needed while maintaining their local cultural integrity.

Group 15: Overcoming obstacles can lead to opportunities for students.

Group 16: Student retention is negatively impacted by a number of variables, which include academic deficiency, financial concerns, family demands, work obligations, motivation, inadequate support systems, and the impact education will have on their lives.

were voted as priorities. Their summary was "Student retention is associated with the institution's academic resources combined with the individual student's preparation and background." For group 6 with 13 groupings, "Students are not adequately prepared to deal with the college transition, as well as institutions are not sufficiently prepared to meet their advising, financial, social and personal needs." For group 9 with seven groupings, "Retention requires strong institutional support systems to overcome a lack of academic preparation and unmet financial need." For group 14 with nine groupings, "American Indian student success is achievable when students are academically prepared, have an educational and career goal and when the institution has the resources to provide the services needed while maintaining their local cultural integrity." For group 16 with seven groupings, "Student retention is negatively impacted by a number of variables, which include: academic deficiency, financial concerns, family demands, work obligations, motivation, inadequate support systems and the impact education will have on their lives."

THE TSU GROUP

Group 8 was composed of ten members from TSU, Malcolm X College, North Carolina A&T, and Southern University. Also note that for group 8, there were only three groupings, which may facilitate the retention planning process because the issues were less complex. This seems to be a function of the fact that these four public Black colleges had so many of the same issues in common related to student underpreparedness. The group 8 workshop opened with an argument on the root cause of the students' problems. Some members felt that student problems were all traceable to poverty, and that even with financial aid students still had an unmet financial need. Others felt that money had nothing to do with a student getting up out of bed and going to class in the morning. At TSU, this is an essential problem—too many students do not come to class. Then the group was in general agreement that many students are academically unprepared and even worse, suffered from low motivation to attain an education. Group 8 was adamant in their opinion that colleges do not have strong academic policies strong enough to overcome prevalent student problems. At TSU, attempts to institute a dress code have so far failed even though it is painfully obvious that many students do not know how to dress properly in school or in the outside world. There is even resistance to establishing dress standard expectations. Furthermore, the group members felt that students need strong academic policies, especially since the student's own support systems tend to be weak. In addition, the lack of strong policies and support systems acts as a deterrent to students and facilitates attrition. Hence, the group 8 summary statement, "Retention requires strong institutional support systems to overcome a lack of academic preparation and unmet financial need." This is the same summary statement reported by group nine.

The Affinity Diagram Process is a widely used management tool (Tague, 2004). One reason why the process seems to work is because it operates much like the general therapeutic process. Individuals come to therapy with a problem and are looking for a quick fix. They may not have the patience to fully analyze the problem, but insights and solutions are best developed only after a full investigation of the problem (Morsund and Kenny, 1993). (Note that of four team TSU members, one did not attend, and one left early because of pressing business. There surely would have been more momentum to the retention plan developed if all four had participated. Nonetheless, the presence of two members from two departments was sufficient to solve certain problems in the initial plan. Group 8 did lose some focus at the point of filling in the retention plan template that may have been due to lack of perceived power to institute the plan.)

THE TSU RETENTION PLAN FOR TEST RESPONSIBLE FRESHMEN REVISED

The last day of the USA Funds Symposium was devoted to mapping out a retention plan using TSU's five-step template. The five questions asked were: (1) What are we going to do? (2) When are we going to do it? (3) Who will be responsible? (4) How much will it cost? and (5) How will we know when it is accomplished?

Table 10.3 presents the revised GUAC/TSU plan for test responsible freshmen. It is directly related to the Group 8 summary statement in speaking to stronger academic policies, academic underpreparation, and perceived financial need, and is also similar in spirit to the original plan. The revised plan has four components. The first is to institute and enforce stronger academic policies by making participation in academic support services mandatory, such as mandatory Study Hall, mandatory tutoring for test responsible freshmen, mandatory participation in the Learning Specialist Program, and mandatory participation in the academic luncheon program. The second is to improve the management of mandatory services as they prepare for an influx of students. The third is to institute mandatory financial aid counseling on budget development and budget planning in order to reduce students' perceived financial need. The fourth is to encourage interactive instructional methods in freshmen classes.

I. INSTITUTE AND ENFORCE STRONGER ACADEMIC POLICIES

Mandatory participation in certain academic support services was part of the original GUAC retention plan, but GUAC did not have the power to enforce it. By going through the USA Funds process, problem-solving abilities must have been enhanced by the concentrated focus on retention, because the group did figure out an approximate strategy. While GUAC cannot mandate participation in academic support services for all students, the department does coordinate Sociology 211, TSU's FYTC, and its requirements. About half of all new students take this course, and most developmental students take it.

There have always been field requirements associated with this course, such as visits to the library and participation in the TSU history project. Other universities have scavenger hunts designed to encourage students to explore and bring back items from various offices and services. Thus, GUAC planned to require ten laboratory hours as part of the grade in which students attend Study Hall and tutoring, visit the Learning Specialist, attend academic luncheons, and one hour of financial aid counseling. Financial aid counseling is explained in more detail below. The academic luncheons are an idea borrowed form Herman Blake and Emily Moore at the University of Iowa (Blake and Moore, 2004). They found that sponsoring brown bag luncheons for Black students that focused only on issues of improving academic performance was extremely successful in retention and particularly successful with Black males. GUAC did not have to worry about mandating class attendance because TSU has already instituted that policy, first with developmental students, and then with all freshman and sophomores. Failure to attend class results in a student being dropped from that class, which may have consequences for financial aid. Sociology 211, the FYTC, is a one-credit course, in which the lab hours count 25% of the grade along with class attendance (25%), tests (25%), and class papers and projects (25%).

Table 10.3   Texas Southern University retention plan*

I. What are we going to do?

1) Institute and enforce stronger academic policies for freshmen by making participation in academic support services mandatory through Sociology 211-Freshman Year Transition Course
   Mandatory class attendance for freshmen and sophomores (already instituted by TSU with penalty of purging.)
   a. Mandatory GUAC Study Hall
   b. Mandatory Bridge Builders Peer Tutoring
   c. Mandatory Learning Specialist Program
   d. 3 Mandatory Academic Luncheons
   e. 1 Mandatory Financial Aid Counseling
2) Improve management of mandatory support services
3) Institute mandatory financial aid counseling, that is, on budget development and planning to reduce the perceived financial need
4) Encourage interactive teaching methods, such as Socratic Method and Supplemental Instruction

II. When are we going to do it?

Phase 1: Fall, 2005
1) First trial, Fall 2005
2) Begin improvement program management and tracking
3) Begin selected counseling for target populations
4) Promote idea to faculty of freshmen and recruit participants

Phase 2: Spring, 2006
1) Second trial, with improvements from fall
2) Evaluate program management and tracking, with improvement plan
3) Widen target population for counseling and evaluate
4) Hold faculty training

III. Who will be responsible?

1) Orientation Committee responsible for communications to freshmen during Orientation. Retention Coordinator (Coordinator of First Year Transition Course), Coordinators of Study Hall. Tutoring, and Learning Specialist Program, and Office of Financial Assistance
2) Program management tracked by Retention Coordinator, Tutoring Coordinator, and Learning Specialist (Advisor Coordinator if First Advisement)
3) Financial aid counseling provided by Financial Aid Office, with tracking accomplished through First Year Transition Courses
4) GUAC in liaison with Provost's Committee

IV. How much will it cost?

1) Academic luncheons only cost. TBD
2) NA
3) Debt counseling is free; Business School offers professional planning; possible internships for business students with GUAC
4) Need grant for faculty incentives; enhanced SI program and more SI leaders

---

(*Continued*)

Table 10.3   (*Continued*)

V. How will we know when it is accomplished?

1) After evaluation of first trial in fall 05. Implementation evaluation determines effectiveness. Effects determined on compliance, attendance, GPA, and retention

2) Data tracking participation in mandatory programs (versus optional in previous years), effects in performance rates, retention rates, student satisfaction rates

3) Number participating and effect on performance rates, retention rates, student satisfaction rates

4) Tracking of percent classes using methods increase, climate ratings increase

*Developed at USA Funds Symposium for Minority-Serving Institutions

This phase of the plan will begin its first trial run in the fall. The Orientation Committee will be responsible for making students aware of Sociology 211 and its requirements during orientation. The Sociology 211 FYTC Coordinator will be responsible for monitoring instructors and helping them with the necessary procedures. Program Coordinators for Study Hall, tutoring, the Learning Specialist, and the Office of Financial Assistance will also be involved. The only cost for this phase is that for academic luncheons.

An evaluation after the first semester will determine the impact of the new laboratory component on student compliance (How many participated in 10 lab hours and what were the GPA consequences?), class attendance, course GPA, semester GPA, and one-year retention. These measures can be compared to similar statistics for the previous year. An implementation evaluation will determine how course management needs to be changed to improve the management of this component. Most important is that GUAC will look for improvement in one-year retention rates for students enrolled in the class compared to the previous year. Having a positive impact on 50% of entering students could make a considerable difference in performance and retention of the freshman class.

II. IMPROVE THE MANAGEMENT OF MANDATORY SUPPORT SERVICES

If more students will be using academic support services, then the job is to make certain that those services are managed efficiently. GUAC hopes for and must be prepared for an influx of students. This will require moving from individually provided services, such as those by tutors, the Learning Specialist, and financial aid, to group services. This will be accomplished by establishing a schedule of weekly group sessions, with varying times each week to accommodate students' schedules. In the case of tutoring, the tutoring staff is relatively small and may easily be overwhelmed. Thus, the plan is to have GUAC lab staff conduct weekly tutorials, also with varying times each week, in math, reading, and writing. However, rather than simply having lab staff available for group sessions, staff will conduct tutorials on certain subjects advertised in advance for the whole semester. That way, students can make better choices in the tutorials they would like to visit. After the group session, lab instructors will make individual appointments. As students attend a service, they sign in and identify themselves as Soc 211 students. At the close of the session they receive a certificate of completion with the date. Ten such certificates will be required.

The new procedures will begin in the fall. The Retention Coordinator (also the Coordinator of Sociology 211) is responsible for making sure that Sociology 211 instructors understand the procedures and acquaint students with them, for overseeing the scheduling of group sessions, and for monitoring student attendance in them. There is no cost involved in the additional monitoring. The student satisfaction surveys will indicate whether this phase has been accomplished and how effective was the process. Student satisfaction surveys are routinely given in group seminars.

III. MANDATORY FINANCIAL AID COUNSELING TO REDUCE PERCEIVED FINANCIAL NEED

Students have enough questions about financial aid to warrant counseling sessions under any circumstances (Foster, 2005). However, these sessions will concentrate on helping students reduce their perceived financial need. According to the Office of Financial Assistance, too many TSU students work excessive hours to pay for cars, apartments, clothes, entertainment, and nonessential expenses. Thus, the heart of the workshop is to assist students in preparing and following realistic budgets. Students will also have access to information and assistance normally provided by the Office of Financial Assistance. This office determines eligibility for financial aid, counsels students on the definition of need, and packages aid. As part of Sociology 211, one financial workshop is mandatory. The same procedures used for other services will be used for this one.

The workshops will begin in the fall. Coordination belongs to the Sociology 211 FYTC Coordinator and the instructors, while the weekly workshops will be conducted by the Office of Financial Assistance. Future plans include enlisting Business School students to conduct debt counseling and financial planning workshops. There is no cost for this service since it can be managed by existing staff. Assessment of accomplishments will involve student participation rates and student satisfaction ratings.

IV. ENCOURAGEMENT OF INTERACTIVE TEACHING METHODS

We have left the encouragement of interactive teaching methods to last, not because it is the least important, but because it is the most difficult to accomplish. These methods fall into two categories: experimental methods already in use by GUAC for which some data has been collected; and existing models and methods that seem promising for future use.

CURRENT EXPERIMENTS IN GUAC

Pre- and Posttesting
One of the simplest and most effective ways of determining course effectiveness is with pre and posttests (Creswell, 2003). This is one of the most important procedures for programs seeking certification by the NADE. Four programs are currently undergoing this process. Sometimes this only means administering a pretest for comparison with

the final exam. In some courses, the content of the pre-post test is clear as in math, but more difficult to target as in an FYTC. This is a first step for any course, and GUAC has employed pre-posttests for a year now in GUAC 101 math, reading, and writing (Adjunct Instructional Programs for college level courses), as well as in Sociology 211, the FYTC. Some results were good, others gave pause.

In all GUAC 101 Adjunct Instructional Courses the posttest scores were higher at a statistically significant level. There was 138.6% improvement in math scores, 23.2% improvement in reading scores, and 21.6% improvement in writing scores. Nonetheless, the posttest scores in math, reading, and writing were all below 70%. Questions are raised such as to why is math more effective, what do reading and writing have to do to increase the rate of improvement, and what must be done to increase posttest scores? That is the work for the coming year.

For Sociology 211, the FYTC, the results were not nearly as good. This one-credit course is designed to teach the study and learning skills necessary for success in college level courses, skills that too many incoming students have not mastered. There is a text for the course. The pre-post measures targeted study habits, academic management skills, and learning styles because previous studies indicated that these skills were important to academic performance (Fleming, 2004). The results showed that there was no positive change on any of the overall measures. There was one subscale for study with people that did show positive change, which is encouraging since Black students tend to be loners according to Fullilove and Treisman (1990). Nonetheless, correlational analysis did confirm that study habits and active experimentation learning styles contributed to better grades.

The problem is that there is a long tradition of instructors for this course teaching it as they please. Some follow the text, others do not. Some focus the course on specialized issues like finance or gender. About half the instructors are GUAC Academic Advisors, and half are other instructors including interns from various graduate school departments. It is the lack of consistency in what is taught, especially when it is ineffective in basic respects, that argues for bringing uniformity in order to deliver a consistent product to students.

For the current year, it is back to the drawing board. An "Academic Management" curriculum has been planned that focuses on teaching critical skills in the context of long- and short- range planning. Instructors not involved in the experimental curriculum will develop individual plans to impart these skills. The effects will be carefully monitored.

Interactive Classrooms in Summer Academy
This Summer Bridge program is designed to better prepare test responsible students for college level work. One of the major goals in Summer Academy is to provide students with a highly interactive classroom climate. Highly interactive classrooms are characterized by procedures that include daily quizzes, use of the Socratic Method of Q&A to establish a dialogue with each student, constructive feedback to students even when their responses are incorrect, and group work with other students. Each instructor has received training in the Socratic Method, The Art of Constructive Feedback, Group Work Scenarios, and TAPPS. In addition, each instructor provides a videotape demonstrating competence in using these procedures. A list of 15 such procedures appears in a Classroom Climate Questionnaire.

In order to determine how effective these procedures are in creating a classroom climate that is perceived by students to be interactive, the Classroom Climate Questionnaire was administered to students at weekly intervals during the Summer Academy program. The average ratings overall and by academic subject were examined for 2003 and 2004 (Fleming 2005a). They indicated that the effort to create more interactive classrooms in Summer Academy appears fruitful in some respects. The results showed that classrooms became more interactive in 2004, compared to 2003, especially in writing, followed by math. The climate in reading classes needed the most improvement. The results also showed that the extent to which students perceive the classroom as interactive has positive consequences for certain performance measures. While there was no relationship to Assessment of Scholastic Skills through Educational Testing (ASSET) placement test scores or pass rates, climate ratings were positively related to program attendance, faculty ratings of student performance, and most importantly, fall GPA and fall credit hours earned. More engaged students may simply be better students, but the Summer Academy classroom procedures are designed to engage more students in the learning process and perhaps in so doing encourage more students to become better.

Multicultural Math
It is often alleged that African American students suffer from a lack of multicultural content in the educational curriculum (Grant and Tate, 2001; Ogbu, 1990). It is further alleged that the presence of culturally relevant content would have a motivating effect, visible in performance and retention (Commissioner's Task Force on Minorities, 1989; Boateng, 1986). While many authors hold this position, there is surprisingly little research that shows whether or not culturally relevant content has a positive effect on the performance of Black students. Indeed, there appear to be no published studies available. Even when instructors do use culturally relevant materials, they do not do so in a way that permits a systematic comparison with standard materials. This effort to enhance the performance of test responsible students (who have not passed all three parts of a state-mandated college entrance test) at TSU has focused on math and reading comprehension.

The subjects for this investigation were a total of 71 test responsible freshmen at TSU enrolled in intensive math and English preparation courses in the Fast Track Program. The students were predominantly African American.

In math, culturally relevant scenarios were created using census data, TSU student data, as well as historical data from sources such as the Civil Rights Movement and the Million-Man March (Fleming et al., in review). Problems were developed on percentages, reading data from graphs (Line graphs, Pie graphs, Histogram), fractions, word problems, and algebra. While there was no pre and posttest, a total of 34 problems were developed, half standard and half culturally relevant. When alternate sets of relevant and standard tests were administered between two classes, students performed up to twice as well (i.e., 100%) on the culturally relevant versions.

In short, the results suggest that the introduction of culturally relevant materials does assist students in learning to do math, and, the culturally relevant information presented was an attempt to teach African American students about their history and the world in which they live. The fact that an influence on performance is also detectable argues for continued efforts in this area.

Reading Comprehension and Multicultural Content
Several previous efforts to develop and use standard and culturally relevant reading comprehension tests show that African American students have a tendency to perform better on culturally relevant exercises and test items. For example, in a Project GRAD Summer Institute for high school students, performance on culturally relevant test items was up to 112% better than on standard items (Fleming, 2002b).

With Texas Southern students, a similar effort during the GUAC Fast Track program for students with low test scores in multiple subjects showed that students performed 100% better on culturally relevant items on pretest, but there was no difference on posttest (Fleming et al., 2004). In this case 12 reading comprehension tests that doubled as full lesson exercises were integrated into the semester's curriculum. The exercises used THEA test-type questioning. There was evidence that students initially performed better on relevant items, but then began to perform better on standard items as the semester progressed. Eventually, there was no difference in performance by the end of the semester on each type of exercise. The tentative conclusion was that the inclusion of relevant items facilitated a transfer of learning to standard items.

In an effort to improve the learning experience for prospective students, a series of eight reading comprehension tests/exercises were used in Summer Academy (Fleming, 2005b). This pilot effort was designed to assess the potential effectiveness of including multicultural test/exercises as part of the curriculum. Note that no pre or posttests were administered, since students were taking the ASSET test at the conclusion of the program. Participants were all Summer Academy students who took reading classes in Session 1 and/or Session 2. There were a total of 78 students in Session 1 and 60 in Session 2. The reading comprehension exercises were previously developed for the GUAC Fast Track Program. This effort with Summer Academy students was only a first attempt, but several conclusions seemed warranted.

1) The use of culturally relevant instructional exercises appeared to improve the performance of male students, but not females.
2) When used with males, performance scores were higher on relevant exercises, and the difference was significant for session 1, but not session 2.
3) Male performance improved on each exercise in the series, as if the male students were learning more with each exercise. Female performance showed less stepwise improvement.
4) Female scores on multicultural exercises were never correlated with ASSET placement test performance, which is a curious phenomenon in itself.
5) Male scores on multicultural exercises showed a number of correlations with ASSET placement test performance, suggesting that the exercises contributed to improving test scores.

Thus, the effort to provide elements of multicultural instruction appears promising, especially for high school students and male students. These initial results warrant continued experimentation.

PROMISING MODELS

Learning Styles and the Teaching of Engineering Students
Richard Felder and his colleagues from North Carolina State University and Stanford University have introduced an experimental instructional approach to training more creative engineers that improves the performance of personality types found in previous studies to be disadvantaged in the engineering curriculum (MBTI extraverts, sensors, and feelers; Felder, 1987, 1993; Felder, Felder, and Dietz, 1998, 2002). The problem, as they saw it, is that engineering requires more creative problem solvers, but engineering school discourages creativity. Given the almost unlimited variety of job descriptions within engineering, Felder (1987) maintains that students with every possible learning style have the potential to succeed as engineers. They may not be likely, however, to succeed in engineering school. The approach then was to use techniques used by educational psychologists to stimulate creative thought. They recognized that to be effective, any new instructional method had to be introduced throughout the curriculum (not in isolated courses), could not take too much class time, and could not take too much of the instructor's time—that is, new methods have to be effective but easy to implement. Felder and colleagues employed the following approaches: (1) The generic quiz, where each week students were asked to make up and solve a take-home final examination, but they had to demonstrate the three higher level thinking skills of Bloom's taxonomy—analysis, synthesis, and evaluation. Using this method, the final examples ranged from good to spectacular; (2) Making up a problem, along the lines of the generic quiz, but on a much smaller scale; (3) Open-ended questions, where students were given a problem and as part of the solution, were required to state what they needed to know to solve it, and how they might go about obtaining the needed information; and (4) Brainstorming exercises, where students were asked to think of as many ways to accomplish a task as possible. After each exercise, the responses were distributed to the class. Each week the number of responses increased steadily. Students performed best on tasks where they had the most practice, but the effect was a visible growth in divergent thinking, an indication that students were being prepared to solve problems (Felder et al., 2002).

Xavier University Model of Science Education
Perhaps the most effective model of innovative change in science teaching is from Xavier University of Louisiana (Sevenair et al., 1987). Student enrollment in the science departments at Xavier was low, which prompted the science faculty to reinvent the way they taught. The key was faculty cooperation and collaboration. Their list of methods is long but includes recruiting in high schools by showing the magical properties of chemistry, establishing a summer program whose goal was nothing short of superstar success in science, and developing customized courses so that all students received the same material regardless of instructor. Xavier is known for using the Whimbey and Lochhead (2000) method of embedding problem solving into the entire science curriculum. Students were taught analytical methods of solving math and word problems, of talking through problem solutions to bring the thinking process to light, and how to analyze their failures in reasoning. The Xavier science faculty published extensively on their teaching experiments. These efforts catapulted

Xavier to the leading edge of science education. Fifty five percent of their undergraduates major in science; 20% of their graduates went to medical or dental school; and Xavier, a small, historically Black institution, put more Black students into engineering than the entire Ivy League combined, according to the American Medical Association.

The 4th Hour Algebra Project

Professor Rene Torres of the University of Texas-Pan American developed the 4th Hour Algebra Project using Title III funds (Torres, 2004). The university's problem with predominantly Mexican American students is similar to TSU's problem: the vast majority of students do not pass college algebra, and without passing it, they cannot graduate. Recruitment of instructors into the project was voluntary, and student enrollment was voluntary. Still, the results soon began to speak for themselves. The ultimate purpose was to instill more uniformity and continuity in all classes within the project and to carefully document the effects of all instructional strategies used, much like the Xavier method. The first principle was increasing time on task. An additional section of the course was added. Quizzes were administered before each test, rather than after so that students would receive more frequent feedback. Supervised group activities were given a central place in instruction, allowing students to problem solve with each other. Grading of homework was eliminated, but students were asked to maintain a course notebook for extra credit. The result was a significant improvement in pass rates in college algebra and a significant decrease in course drop rates (Torres, 2004).

The Deliberate Teaching of Critical Thinking

An unpublished project at Barnard College in the Psychology Department experimented with the deliberate teaching of critical thinking. Higher order critical thought, such as comparative analysis and analysis of argument, is a byproduct of good education, but is almost never taught directly and deliberately (Resnick, 1987; Kurfiss, 1988). This project adapted a method of scoring compare and contrast essays that was based on the research of McClelland, Winter, and Stewart (1983). The scoring method was converted to an instructional method where instructors taught students to write good compare and contrast analysis essays. Furthermore, during a second phase of the project the course curriculum was converted to compare/contrast format. Thus, each topic posed a main question in comparative terms, and every exam and paper was recast in compare/contrast terms. Early attempts to teach the method on a limited basis were less successful than when the whole course was cast as a comparative analysis course. The message was that students gain skill when they are given ample opportunity to practice. The outcome was that student exams improved noticeably, but that student term papers became so sophisticated that a number were of publishable quality. The course also stimulated student interests in pursuing research, which greatly pleased the faculty.

With underprepared students, the level of teaching effectiveness must be raised. The laissez-faire prerogative that faculty have come to expect may need to be rethought. Evidence that faculty cooperative in setting joint strategic goals to improve student performance in critical ways may have a higher priority than academic autonomy. There is no evidence that secondary school preparation is improving the quality of incoming students at colleges like TSU, and there are in fact assurances from state educational officials, such as Commissioner of Higher Education Raymond Paredes, that no such improvement can be expected. Therefore, the best course of action may be to recalibrate the process of instruction, instruction carefully calculated to rescue student talents.

CONCLUSION

The retention planning process at TSU is an evolving process that is continually updated, revised, and evaluated. Sometimes, a plan encounters circumstances that threaten to derail it; other times, events act to facilitate or energize the plan. One such energizing event was participation of TSU at the USA Funds Symposium for Minority-Serving Institutions. As a financial institution concerned about the loan default rate among students, they sponsor this symposium to assist institutions with refining retention plans that would help keep students in school and minimize the default rate in accordance with the retention literature. Instructional initiatives designed to enhance student performance constitute the heart of the retention plan in the GUAC. However, the essential problem for the GUAC has been the perceived lack of power to implement promising retention strategies. The USA Funds symposium played a critical role in assisting the problem solving process by facilitating interdepartmental cooperation, introducing an effective method of analyzing problems, utilizing collegial group work, providing supporting information, and showcasing effective models. For GUAC, the result was a revised four-part retention plan with no perceived barriers to implementation that entailed interdepartmental cooperation. According to recent literature, low performance is a generalized problem for Black students in any college setting (Astin et al., 1996; Galloway and Swail, 1999), but performance appears to be the real key to keeping Black students in Black colleges.

NOTE

1. The 15 steps in the Affinity Diagram Process are as follows: Prepare; Explain the purpose of the exercise; Have an open discussion; Generate ideas; Clarify the ideas; Review for missing ideas; Group the ideas; Generate titles for the groups; Show the relationship among groupings; Make a layout of the groupings; Establish priorities among the groupings with points; Total the points; Summarize the outcome; Own it (participants to sign and date the diagram); Reflect on the process.

REFERENCES

Astin, A. W., L. Tsui, and J. Avalos. 1996. *Degree attainment rates at American colleges and universities: Effects of race, gender, and institutional type.* Los Angeles: Higher Education Research Institute, University of California at Los Angeles.

Blake, H. J. and E. L. Moore. 2004. Retention and graduation of Black students: A comprehensive strategy. In *Best practices for access and retention in higher education,* ed. I. M. Duranczyk, J. L. Higbee, and D. B. Lundell, 63–71. Minneapolis, MN: Center for

Research on Developmental Education and Urban Literacy, General College, University of Minnesota.

Boateng, F. F. 1986. Multicultural education in a monocultural classroom. *Viewpoint: Journal on Teaching and Learning* 6: 2–4.

Commissioner's Task Force on Minorities. 1989. *A curriculum of inclusion.* New York: Author.

Creswell, J. W. 2003. *Research design: Qualitative, quantitative and mixed method approaches.* Thousand Oaks, CA: Sage.

DeSousa, J. 2001. Reexamining the educational pipeline for African American students. In *Retaining African Americans in higher education,* ed. L. Jones, 21–44. Sterling, VA: Stylus.

Felder, R. M. 1987. On creating creative engineers. *Engineering Education* 78: 222–7.

———. 1993. Reaching the second tier: Learning and teaching styles in college science education. *Journal of College Science Teaching* 23(3): 286–90.

Felder, R. M., G. N. Felder, and E. J. Dietz. 1998. A longitudinal study of engineering students performance and retention. V. Comparisons with traditionally-taught students. *Journal of Engineering Education* 87: 469–80.

———. 2002. The effects of personality type on engineering student performance and attitudes. *Journal of Engineering Education* 91(1): 3–17.

Fleming, J. 1984. *Blacks in college.* San Francisco, CA: Jossey-Bass.

Fleming, J. 2002a. Who will succeed in college? When the SAT predicts Black student's performance. *Review of Higher Education* 25(3): 281–96.

Fleming, J. 2002b. *The effects of culturally relevant content on reading comprehension in low achieving students: Implications of the "No Child Left Behind" policy on historically Black colleges.* Paper presented at the conference on African American Education in the South: "No Child Left Behind," Historical and critical analysis of current educational policies on African American students' achievement. Dillard University and Longue Vue House & Gardens, New Orleans, November 15. Available from http://fleming_jx@tsu. edu or http://JacquelineFleming@yahoo.com.

Fleming, J. 2004. *When under-prepared students stay in school: The Fast Track Program at Texas Southern University.* Unpublished manuscript, General University Academic Center, Texas Southern University, March. Available from http://fleming_jx@tsu.edu or http://JacquelineFleming@yahoo.com.

Fleming, J. 2005a. *Final Report: Summer Academy 2004: Classroom climate.* Report Supplement I. General University Academic Center, Texas Southern University, May.

Fleming, J. 2005b. *Final Report: Summer Academy 2004: Multicultural content and reading comprehension.* Report Supplement II. General University Academic Center, Texas Southern University, May.

Fleming J., J. Guo, J. Howard, D. Lewis, T. Stroud, and S. Zamora. 2004. *The power of attendance as a retention tool: Attendance, performance and retention of Texas Southern University freshmen.* Paper presented at the General University Academic Center (GUAC) Open Advisory Meeting on Student Retention, Texas Southern University, November 11. Available from http://fleming_jx@tsu.edu or http://JacquelineFleming @yahoo.com.

Fleming, J., J. Guo, S. Mahmood, and C. R. Gooden. 2004. Effects of multicultural content on reading performance. In *Best practices for access and retention in higher education,* ed. I. M. Duranczyk, J. L. Higbee, and D. B. Lundell, 55–62. Minneapolis, MN: Center for Research on Developmental Education and Urban Literacy, General College, University of Minnesota.

Fleming, J., J. Guo, S. Mahmood, and C. R. Gooden. in review. Effects of multicultural content on mathematics performance. *Journal of Black School Educators.*

Fleming, J., D. Maddox, V. McReynolds, and C. Wilder. 2005. *Retention and Performance of First Time Freshmen at Texas Southern University: 1993 to 2003.* Unpublished paper, Texas Southern University. Available from http://fleming_jx@tsu.edu or http://JacquelineFleming@yahoo.com.

Foster, C. 2005. *Financial literacy for teens.* Conyers, GA: Rising Books.

Friedman, P., F. Rodriguez, and J. McComb. 2001. Why students do and do not attend classes. *College Teaching* 49(4): 124–33.

Fullilove, R. E., and P. U. Treisman. 1990. Mathematics achievement among African American undergraduates at the University of California, Berkeley: An evaluation of the mathematics workshop program. *Journal of Negro Education* 59(3): 463–78.

Galloway, F. J., and W. S. Swail. 1999. *Institutional retention strategies at historically Black colleges and universities and their effects on cohort default rates: 1987–1995.* Washington, DC: Sallie Mae Education Institute. Also available from http://www.educationalpolicy.org.

Grant, C. A., and W. F. Tate. 2001. Multicultural education through the lens of the multicultural education research literature. In *Handbook of research on multicultural education,* ed. J. A. Banks and C. A. M. Banks, 145–68. San Francisco, CA: Jossey-Bass.

Kurfiss, J. G. 1988. *Critical thinking: Theory, research, practice and possibilities.* ASHE-ERIC Higher Education Report No. 2. Washington, DC: Association for the Study of Higher Education.

McClelland, D. C., D. G. Winter, and A. J. Stewart. 1983. *A new case for the liberal arts.* San Francisco, CA: Jossey-Bass.

Morning, C., and J. Fleming. 1994. Project preserve: A program to retain minorities in engineering. *Journal of Engineering Education* 83(3): 1–6.

Morsund, J., and M. C. Kenny. 1993. The process of counseling and therapy. Englewood Cliffs, NJ: Prentice-Hall.

Mow, S. L., and M. T. Nettles. 1990. Minority access to and persistence and performance in college: A review of trends in the literature. In *Higher education: Handbook of theory and research,* vol. 6, ed. J. Smith, 35–105. New York: Agathon.

Ogbu, J. 1990. Literacy and schooling in subordinate cultures: The case of Black Americans. In *Going to school: The African American experience,* ed. K. Lomotey, 113–31. Albany, NY: State University of New York Press.

Pascarella, T., and P. Terenzini. 1977. Patterns of student-faculty attrition. *Journal of Higher Education* 48(5): 541–52.

Porter, O. 1990. *Undergraduate completion and persistence in four-year colleges.* Washington, DC: National Institute of Independent Colleges and Universities.

Resnick, L. B. 1987. *Education and learning to think.* National Research Council, Washington, DC: National Academy Press.

Richardson, R. C., and L. W. Bender. 1987. *Fostering Minority Access and Achievement in Higher Education: The Role of Urban Community Colleges and Universities.* San Francisco, CA: Jossey-Bass.

Sevenair, J. P., J. W. Carmichael, J. Bauer, J. T. Hunter, D. Labat, H. Vincent, and L.W. Jones. 1987. SERG: A model for colleges without graduate programs. *Journal of College Science Teaching* 16(5): 444–6.

Sherman,T. M., M. B. Giles, and J. Williams. 1994. Assessment and retention of Black students in higher education. *Journal of Negro Education* 63(2): 164–80.

Tague, N. R. 2004. *The quality toolbox.* Milwaukee, WI: Quality Press.

Thomas, P. V., and J. L. Higbee. 2000. The relationship between involvement and success in developmental algebra. *Journal of College Reading and Learning* 30(2): 222–32.

Tinto, V. 1975. Dropout from higher education: A theoretical synthesis of recent research. *Review of Educational Research* 45: 89–125.

———. 1987. Dropping out and other forms of withdrawal from college. In *Increasing student retention,* ed. L. Noel, R. Levitz, D. Saluri, and associates. San Francisco, CA: Jossey-Bass.

Torres, J. R. 2004. *The fourth hour college algebra project: Improving student retention.* Paper presented at the General University Academic Center (GUAC) Open Advisory Meeting on Student Retention, Texas Southern University, November 11. Available from http://jrtorres@panam.edu.

UNCF Statistical Report. 1998. Fairfax, VA: United Negro College Fund.

Whimbey, A., and J. Lochhead. 2000. *Problem solving and comprehension.* Hillsdale, NJ: Lawrence Erlbaum.

PART 3

TABOOS

CHAPTER 11

COLOR AND CLASS: THE PROMULGATION OF ELITIST ATTITUDES AT BLACK COLLEGES

BIANCA TAYLOR

> Wouldn't they be surprised when one day I woke out of my black ugly dream, and my real hair, which was long and blond, would take the place of the kinky mass Momma wouldn't let me straighten? My light-blue eyes would hypnotize them.
>
> Maya Angelou

This historical analysis uses the Black college as a lens through which to explore the often skirted issues of classism and colorism within the Black community. Blacks living in America have been the targets of racism since the inception of slavery; it is well known that the origin of much of the intolerance sat within the White community. However, arguably less widely recognized is the intraracial prejudice rooted in the Black community.

This chapter follows the dynamics of internal racism and the stratifying agent, class, within the Black community from their origin during the days of slavery and traces their evolution through the Black college. Within the Black community from the antebellum era to the Black Power Movement, those with fairer complexions and established family status were afforded greater opportunity. Such preferential treatment drove a wedge between dark-skinned and fair-skinned Blacks. This phenomenon was perhaps most evident within historically Black colleges and universities (HBCUs). In this chapter, the Black college is established as a polarizing institution that serves as a distillery within which these phenomena are more easily perceivable.

HBCUs in the United States have played a significant role in the socioeconomic status of Black men and women, as education has been the principal social factor responsible for the emergence of the Black bourgeoisie (Frazier, 1957; Gatewood,

2000). This chapter will examine elitist attitudes at Black colleges and how a critical dichotomy based on complexion and ultimately class has haunted the Black community from the days of slavery to present.

LITERATURE REVIEW

There are a number of scholars who address the issues of color and class. Russell, Wilson, and Hall's (1990) *Color Complex* proved to be an enlightening and comprehensive take on the subject matter. This book follows the debilitating issue of colorism from slavery to present day tribulations in the professional and personal life of the Black individual. Although there is some literature that pertains to Black upper class and skin color, there is limited information on colorism and classism at HBCUs. The majority of the available research that references the education of the Black elite focuses on the free, fair-skinned Blacks who attended predominantly White colleges in the North and Northeast that afforded some Black students admission (Frazier, 1957; Gatewood 2000). Information on the Black elite attending HBCUs is scarce, with the exception of a few sources. Social scientist and scholar E. Franklin Frazier, public intellectual Lawrence O. Graham, and historian Willard B. Gatewood offer valuable insight into the lives and the experiences of the Black elite in *Black Bourgeoisie* (1957), *Our Kind of People: Inside America's Black Upper Class* (2000), and *Aristocrats of Color: The Black Elite, 1880–1920* (2000), respectively. In *Black Bourgeoisie*, Frazier details his controversial study of the privileged status of the Black middle-class, more specifically mulattoes and light-skinned Blacks. He argues that Blacks with fair skin were more likely to be emancipated during slavery and to receive an education. Frazier reasons that one's skin tone is the stratifying agent that determines one's access to privilege and the Black middle-class or the Black bourgeoisie. Frazier often casts Black education in a negative light, stating, "From its inception the education of the Negro was shaped by bourgeois ideals" (1957, p. 60). He implies that the Black bourgeois' objective was to use a college education for personal benefit, and not to uplift the Black race. The underlying message is that the assimilation philosophy of the Black middle class will ultimately destroy the race. In *Our Kind of People* (2000), Graham concentrates on the socially acceptable colleges for the Black elite; these were HBCUs that stressed a liberal arts curriculum and attracted Blacks with fair skin and middle to high incomes. *Our Kind of People* also focused on vacation spots, social clubs, and Greek organizations exclusive to the Black elite. Gatewood's *Aristocrats of Color* chronicles the lives of the most prominent families, often defined by one's credentials, wealth, accomplishments, and color. In this study of the Black upper class immediately following Reconstruction, Gatewood details their struggle with identity as they advocate for the rights of Blacks while physically removing themselves from the "Black masses" and in some cases, longing for the approval of Whites.

THE CREATION OF CLASSISM AND COLORISM DURING SLAVERY

According to the Center for Race and Gender, "colorism is in effect when one's complexion becomes the basis for awarding, restricting or denying access to power and resources in various arenas of society. Such discrimination produces a skin tone hierarchy (http://crg.berkeley.edu/programs/programs)." Racial supremacy, mostly White supremacy, informs the principle of colorism. Russell, Wilson, and Hall (1993, p. 80) acknowledge the absurdities of intraracial skin-color discrimination in *The Color Complex:*

> Nowhere else in the world does a single race encompass people whose skin color ranges from white to black, whose hair texture varies from tightly curled to straight, and whose facial features reflect the broadest possible diversity. Were it not for this artificial grouping, part of the legacy of racism, blacks might not criticize each other so harshly for having skin or hair the does not meet some arbitrary standard.

Colorism and classism have a long, ugly history within the African American community, dating back to slavery when White slaveholders created a caste system among slaves based on complexion. However, colorism took root within the Black community when they accepted and exploited this contrived hierarchy (Harvey et al., 2005). Since the commencement of slavery, Whiteness was associated with all that was civilized, virtuous, and beautiful, while Blackness was identified with being untamed, sinful, and revolting (Hill, 2002). Colorism within the African American community stems from the miscegenation between White men and Black female slaves, and occasionally between White women and Black male slaves or freedman (Russell, Wilson, and Hall, 1993). In the case of the White men and the Black female slaves, often the slaves, who were typically of a dark complexion, were raped by their White slave owners (Keith and Herring, 1991). Nevertheless, the result of many of these illegitimate affairs was the birth of biracial or mulatto children. As such rendezvous were taboo, rarely did the slave master accept his child(ren) in his home as his own. Despite considering their illegitimate children to be slaves, they gave them special treatment by allowing them to work in the house, as opposed to toiling in the field (Du Bois, 1903; Frazier, 1957; Keith and Herring, 1991). Du Bois (1903, p. 552) addressed this phenomenon by stating that "some were natural sons of unnatural fathers and were given often a liberal training and thus a race of educated mulattoes sprang up to plead for black men's rights."

Despite being considered Negro or of a lesser race by the slave master, fair-skinned mulattos were referred to as delicate and, therefore, made worthy of being a house servant because of their European-like physical appearances (Johnson, 1996; Kerr, 2005). In fact, many slaveholders insisted, "no man would buy a mulatto for field work" (Johnson, 1996, p. 111). Those with fair skin were considered to be fit for intelligent tasks, while those with dark skin were considered "healthier" and better suited for manual labor (Kerr, 2005; Toplin, 1979). As mentioned in a 1947 issue of *Ebony* magazine,

> Mulattoes brought big prices in the slave market. So Maryland enterprisingly bought out a "fancy" line of mulatto stock . . . Undoubtedly, licentiousness among the white breeders played no small part in the mulatto production . . . Good-looking mulatto girls were in demand among the young men without families, who were making the lonely trek to the western frontier.
>
> (May 1947, p. 33)

The mulattoes who received their freedom papers contributed to the foundation of the free Negro population that flourished in the South before the Civil War

(Toplin, 1979). According to Frazier (1957), in 1850, mulattoes constituted 37% of the free Negro population, but only 8% of the slave population.

The fair-skinned slaves typically worked and lived in the slave master's house, and overwhelmingly they were selected for more desirable jobs, such as housework, caring for the master's White offspring, and the position of field foreman. As field foreman, fair-skinned slaves were given authoritarian position over dark-skinned slaves. Alternatively, most dark-skinned Black slaves were forced into physical labor, sometimes under the fair-skinned slaves' orders (Frazier, 1957; Keith and Herring, 1991). Dark-skinned slaves regularly worked on the harvest, picked cotton, built structures, and smelted iron. Because of the betrayal of and lack of sympathy for dark-skinned field slaves by some of the fair-skinned house slaves, dark-skinned slaves quickly learned that their kin was not always their brother. The fair-skinned Blacks who abandoned their race were considered Uncle Toms or turncoats. Benjamin Hudson (1963, p. 79) provides the following definition in his *Another View of "Uncle Tom"*:

> The name "Uncle Tom" has become . . . synonymous with everything that is . . . cowardly and contemptible . . . used to designate a person who, through fear or desire for personal gain, betrays the trust of those whom he represents, who acquiesces to the wishes and dictates of a more powerful group, who is generally without scruples or principles, and who is always lacking in moral courage.

Not only did fair-skinned slaves reap benefits because of their skin color, but some were also able to circumvent slavery completely because they were able to "pass" as being White (Toplin, 1979). When faced with the question of a lifetime of slavery and heartache or freedom and possible upward mobility, countless fair-skinned mulattoes, quadroons (a person with one-quarter Black or African heritage and three-fourths White or European heritage), and octoroons (a person who is seven-eighth White and one-eighth Black) chose freedom and essentially lived their lives as a White person (Bennett, 2001).

EDUCATION OF SLAVES

Not only did many fair-skinned slaves experience the benefits of working in the master's house, but many were taught to read, given an education, provided with apprenticeships, and exposed to wealthy White traditions (Frazier, 1957; Keith and Herring, 1991; Kephart, 1948). According to Toplin (1979), mulatto slaves were more likely to receive an education and their freedom papers; and in at least one instance, a freed mulatto was given his own slaves and plantation.[1] An educated slave was an anomaly during the slavery era, as it was illegal for slaves to read or be educated because it was seen "as a threat to their [the White majority] ability to control and manipulate enslaved Africans" (Williams, 2004, p. 8). However, a number of slaveholders educated some of their slaves as it was economically advantageous; moreover, many fair-skinned slaves used this education and their adopted-White values to assimilate into White America and to draw a sharp line between themselves and their dark-skinned brethren (Frazier, 1957; Gatewood, 2000; Kephart, 1948; Russell, Wilson, and Hall, 1993). To prevent being deemed a "bad Negro" (Gatewood, 2000) or associated with the poorer, darker-skinned newly freed slaves by Whites, many mulattoes exclusively socialized with mulattoes and actively discriminated against those darker than them (Jones, 2000).

Such a superiority complex likely placed a negative self-image upon those of a darker complexion and those who were forced into manual labor (Golden, 2004). Du Bois (2005) pioneered this idea in his theory of double consciousness, which proposes that Blacks internalized the perspective of Whites. And while this was nothing but detrimental to all of society, those of the darkest hue were the greatest victims. The slaveholders' creation of a caste system or a social hierarchy amongst slaves, as well as the entire population, fostered the proliferation of internal racism or colorism and eventually classism throughout the Black community (Golden, 2004; Frazier, 1957; Kerr, 2005). Although it may not have been the initial intention of the slaveholder, the division amongst slaves and between slaves and freedman based upon melanin created a fatal dichotomy amongst Blacks (Toplin, 1979). This division forced them to turn on one another, as opposed to coming together to confront the slave master or hate mongering Whites (Toplin, 1979). Not all historians, Sterling Stuckey for example, believe deep divisions existed between the house and field slaves. Stuckey (1987) supports the theory that the interactions and similarities between all of the slaves created a single culture. He suggests that allegiance to Africa proscribed Christian influence and American alliance, therefore, created a homogeneous slave culture.

According to Frazier (1957, pp. 12–3), "generally, the son of a house servant was apprenticed to some artisan to learn a skilled trade. These skilled mechanics, who constituted a large section of the artisans in the South, formed with the house servants a sort of privileged class in the slave community." Incorporated in this Negro education were the traditions and values of New England and the Puritans (Frazier, 1957). Consequently, through the promotion of White New England values, Blacks were conditioned to appreciate all things White and Northern, and not to act on behalf of the interests of Black people.

PROBLEMS WITH W. E. B. DU BOIS'S TALENTED TENTH

According to Du Bois, "the Negro race, like all races, would be . . . saved by its exceptional men. The problem of education . . . among Negroes must first of all deal with the Talented Tenth; it is the problem of developing the Best of this race that they may guide the Mass away from the contamination and death of the worst, in their own and other races" (Frazier, 1957, p. 68). Alternatively, William S. Scarborough, a Black scholar and former president of Wilberforce University, supported the concept of Black aristocrats helping the less fortunate Negroes, but by distancing themselves from Blacks of lower socioeconomic status. He argued that the elite Blacks should "form classes of society where culture and refinement, high thinking and high living, in its proper sense, draw the line" (Gatewood, 2000, p. 118). However, in *The Negro Problem*, Du Bois (1903, p. 553) states that "the Talented Tenth rises and pulls all that are worth the saving up to their vantage ground." Regrettably, too many of the Black elite followed Scarborough's ideals and abandoned the "submerged masses," and heeded Du Bois' words literally by pulling only those "worth the saving" (Du Bois, 1903, p. 553).

The notion of *us* versus *them* became common mindset as many affluent and influential Negroes referred to the less fortunate as "those niggers." Graham's *Our Kind of People: Inside America's Black Upper Class* (2000) describes his great-grandmother's emphasis on complexion and her distinction between *them* and *us*. According to Graham (2000, p. 2), his great-grandmother would say "'Niggers, niggers, niggers.' She would say [this] under her breath while staring at the oversized pages . . . of Negro politicians, entertainers, and sport figures who were busy making black news in 1968." Graham (2000, p. 3) recalled his great-grandmother being "proud when a black man finally won an Academy Award, but was disappointed that Sidney Poitier seemed so dark and wet with perspiration when he was interviewed after receiving the honor." Although the Black elite had an *us*-versus-*them* mindset, many in White America saw no distinction. Despite the Black elite's efforts to drastically detach themselves from the "submerged masses," White America still considered them to be Negro, and thus inferior (Gatewood, 2000). As stated in *Ebony* magazine (1945, p. 3), "In the race-riddled pattern of American life, color has heaped together all men with black skins, regardless of brains or brawn."

ELITE SOCIETIES AND SOCIAL CLUBS: THE BROWN PAPER BAG TEST

The color complex was not class-blind; issues of color plagued those with no formal education to the respected community doctor. However, Golden (2004) supports the notion that the Black aristocracy "developed more openly elaborate rituals, attitudes, and strategies to maintain lightness, White blood, and the status it conferred the other Blacks" (p. 43). According to Gatewood, "For the colored aristocrats . . . nothing was more absurd than the idea that all Blacks were social equals. They [the Black elite] viewed such notions as utter fiction, based largely on the man's ignorance of the black community and positively detrimental to racial progress" (2000, p. 113). As a result, many wealthy and fair-skinned Blacks used things like education and elite societies to create more distinct boundaries between themselves and dark-skinned and/or low-income Blacks. An article from *Ebony* (1945, p. 3) supports this notion:

> Only one of every 130 Negroes in America went to college . . . [Negro college graduates] know that perhaps in nation or race is the gap between the polished college graduate and the illiterate laborer so narrow as between the one Negro university alumnus and the 129 who never saw a campus.

After emancipation, many fair-skinned Blacks used their education and their adopted-White values to assimilate into White America, as much as White America would accept them (Jones, 2000). Fair-skinned Blacks and mulattoes were more likely to gain well-paying employment, receive a formal education, travel internationally, and be moderately accepted by Whites in the general population (Toplin, 1979). In addition, the benefits given to the generations of fair-skinned slaves gave their successors a significant advantage over dark-skinned freedman in creating a "normal life." Monopolizing on their education, fair complexion, and financial status, the Black elite created exclusive clubs or "blue vein" societies and based admission criteria on skin color, European-like features, wealth, and family status (Graham, 2000).

The most popular method of determining acceptance was the brown paper bag test; in order to gain membership into Black elite societies one's skin had to be lighter than a brown paper bag (Maddox, 2002). Not only did societies and organizations use the paper bag test, churches, secondary schools, and even barbershops used complexion tests to determine social acceptance and social status (Golden, 2004; Kerr, 2005). The methods used to determine admission into the elite Black societies were accepted and common knowledge in the Black community during the nineteenth century and part of the twentieth century (Graham, 2000). Other methods or tests used to determine membership included the comb test, ruler test, and the blue vein test (Russell et al., 1993). The comb test consisted of running a fine-tooth comb through a person's hair. If there was any resistance, the person would be considered "too Black" and denied membership (Graham, 2000). The ruler test consisted of placing a ruler next to a prospective member's hair; if their hair was not as straight as a ruler was, they were not admissible. In addition, the blue vein test was another method of determining one's White heritage; the prospective member's blue veins had to be visible through their fair skin (Graham, 2000). Such complexion admission tests for elite societies show how some elite fair-skinned Blacks of the time valued European-like features and sought to emulate Whites physically, socially, and economically. From the viewpoint of the Black elite, successfully passing the complexion tests were verification that they were fair enough to be removed from the negative stereotypes of dark-skin Blacks by exaggerating their physical likeness and compliance to White America.

Despite rejection from the greater White community, there were members of the Black upper class who still refused to embrace those of lower-income and those with dark complexions within their own race; thus, the promulgation of blue vein societies. According to Toplin (1979), fair-skinned freedmen created the Brown Fellowship Society, which forbade interaction with, and marriage to, dark-skinned Blacks. Being a member of the Links Incorporated, the Girl Friends, the Northeasterners, or the National Smart Set was the epitome of Black women of high society. Meanwhile, the men of the Black elite endeavored to join such organizations as the Boulé (Sigma Pi Phi), the Comus Club, the Guardsmen, and the One Hundred Black Men. And Jack and Jill is a by-invitation-only organization for children of the upper echelon of Black high society (Graham, 2000). While many of these clubs and organizations have a history of elitism based on color and class, they "were founded on a premise of volunteerism and charitable giving . . . " (Gasman and Anderson-Thompkins, 2003, p. 17). In addition, well-to-do Blacks attempted to further remove themselves from any interaction with the Black masses by moving into exclusive neighborhoods such as Hyde Park in Chicago, LeDroit Park and the "Gold Coast" in Washington, DC, Strivers Row in Harlem, downtown New Orleans and Mount Airy in Philadelphia (Golden, 2004; Graham, 2000). According to Golden, Washington, DC was the "capital" of colorism, as it was home to a flourishing group of freed mulattos (2004).

Before elite Blacks used racial identity or complexion tests, Whites used "scientific" tests to prevent fair-skinned Blacks from "passing" as White (Kerr, 2005; Larsen, 1969). Some of the purported "scientific" tests used by Whites to determine a person's race and ethnicity include examining the color of one's nail bed, palms, shapes of ears, teeth

(Kerr, 2005; Larsen, 1969, p. 150). As early as the African slave trade, many Whites believed the Negro race to be inferior (Gatewood, 2000; Wesley, 1940). The notions of inferiority were founded in the racial differences, such as skin color, facial features, and inherited traits (Wesley, 1940). Therefore, "scientific" tests served as an extreme manner by which to establish a hierarchy between Negroes and "superior" Whites.[2]

Those with both White and African lineage were identified as "tragic" and "mongrel" (Gatewood, 2000). Many in mainstream White America refused to accept any mulatto or mixed race person as White, no matter how fair their complexion; and the defining test used to differentiate Whites from Blacks was the one-drop rule (Hickman, 1997). The one-drop rule declared that anyone with one drop of Black blood was of the Black race and had "the same legal status as a pure African" (Jones, 2000, p. 1505). For example, a person who was a quadroon or an octoroon was considered fully Black.

EARLY BLACK COLLEGES

By the end of Reconstruction, education served to buttress the social construct of the Black elite. For the benefit of social mobility, many Black elites continued to differentiate themselves from the "submerged masses" by enrolling in predominantly White colleges that claimed to draw no color line. "From the perspective of the fair-skinned, status-minded parents, attending all-white colleges and universities was quickly becoming a calling card of the black elite, and attending a white college preparatory school would secure entrance into Harvard, Yale, or Oberlin, the choice schools of the black elite" (Kerr, 2005, p. 276).[3] According to John Thelin (2004), many Black women were erroneously offered admission to elite colleges because they were able to "pass" as White. Alternatively, many elite Blacks enrolled at what is known today as the Negro Ivy League. Similar to many prestigious private White institutions in the Northeast, schools in the Negro Ivy League enroll third- and fourth-generation legacies in the present day (Graham, 2000). According to Graham (2000, p. 66),

> just as the Roosevelts and the Kennedys had Harvard and the Buckleys, and the Basses had Yale, old families among the black elite have selected certain colleges for their children and their descendants. While northern blacks and some free southern blacks certainly chose to come north during the 1800s to attend white colleges and universities like Amherst, Harvard, or Oberlin College in Ohio, most members of the black elite attending college during the immediate post-Civil War period preferred to establish their family roots at the black southern universities founded by religious organizations.

While the number and nature of HBCUs are vast, only a select few HBCUs are considered members of the Negro Ivy League. Unlike the Ivy League, the Negro Ivy League is not an official set of institutions or an athletic division; it is a subjective grouping of the "best" Black colleges. Therefore, identifying which colleges are and are not included in the group may vary according to perspective. The classification of the Negro Ivy League is often based upon prominent alumni in various fields, highly credentialed faculty members, and the social status of its students (Graham, 2000). For some, the Negro Ivy League includes Dillard University, Fisk University, Hampton

University, Morehouse University, Spelman College, and Tuskegee University (Gasman, 2006). However, others, such as public intellectual Graham (2000), consider Howard University, Morehouse, and Spelman to be the crème de la crème of higher education for the Black elite, as the majority of institutions have a liberal arts curriculum, are highly respected in the White professional world, and their graduates gain admission to competitive graduate schools and boast highly regarded professions. Graham also recognizes Bennett College, Clark-Atlanta University, Fisk, Hampton, Lincoln University, Meharry Medical School, Tuskegee University, and Xavier University as influential institutions in the making of the Black upper class (2000).

In 1945, *Ebony* referred to the top Negro colleges of the time as the "Big Four"— Howard University, Atlanta University, Lincoln University, and Fisk University. *Ebony* (1945, p. 3) described one of the schools in the following manner: "Perhaps the crucial key to this Maginot Line of Negro education is historic Fisk University; traditionally ranked as the Yale of Negro colleges . . . For the students it meant standards as stiff as the topmost white schools." With limited access to historically White institutions (HWIs) prior to the *Brown v. the Board of Education* decision, Black elites often considered Howard the "Black Harvard." Moreover, Spelman College often competed with the esteemed Seven Sisters colleges[4] for talented Black female students (Graham, 2000, p. 75).

From the late nineteenth to mid-twentieth century, HBCUs served as either a barricade or a conduit toward achieving social acceptance and influence within the Black community. Whether Black colleges were a barrier or a conduit to success depended upon one's skin color, family status, and the education level of the family (Gatewood, 2000). Social class and family lineage often dictated one's social network, which was directly correlated to employment opportunities, connections to the affluent and influential and resources that in general improved quality of life. Lura Beam, a White teacher at Black schools in the South, was dismayed to learn that notwithstanding the prejudice and intolerance against Blacks by Whites, the Black aristocrats only championed the prejudice by creating more avenues for such against other Blacks. According to Beam (Gatewood, 2000, p. 114),

> Among aristocrats of color in the early twentieth-century South, a man without a distinguished family background and of dark complexion had no chance of being accepted into the highest social circles unless he possessed an advanced degree from a prestigious northern university, in which case "adjustments would be made." For example, the Bond family of Kentucky, though fair in complexion, lack most of the attributes required for membership in the aristocracy of color except that of education. The Bonds possessed neither a distinguished old family background nor wealth, but the educational achievements begun late in the nineteenth century by Henry and James Bond, and continued by their descendants, gained for the family a place in the upper stratum of Black society.

Therefore, higher education, specifically HBCUs, contributed to the creation of the Black bourgeoisie and made the separation between the classes or castes more distinctive. The social networks created in colleges helped to perpetuate an advantage that continued after the college years. Many of these institutions originally were considered "cultural starting schools" or places where students began the accumulation of social connections and support that would not only ease their transition and progression through college, but also increase their social market value and employment opportunities after graduation, similar to the networking at White elite institutions (Gatewood, 2000).

The change in the mission of Black colleges also contributed to the creation of the Black bourgeoisie. Frazier (1957) supports the notion that the purpose or objective of Black colleges shifted during the first half of the twentieth century from honing Black intellectuals and uplifting the community to a focus on producing wealth and materialism. According to Frazier (1957, p. 76), a top administrator at Tuskegee Institute concluded, "A man never begins to have self-respect until he owns a home." While many disagree with the provocative views of Frazier, some like Langston Hughes, also critiqued higher education. James D. Anderson (1988, p. 212) notes, "In 1934, writer and poet Langston Hughes denounced the 'cowards from the colleges,' the 'meek professors and well-paid presidents,' who submitted willingly to racism and the general subordination of black people." Many Black college teachers also adopted this shift in objectives. Frazier (1957) claimed Black teachers often went into the teaching profession for the social and economic benefits, while White teachers at HBCUs were more interested in racial justice for all. Similar to physicians, dentists, and lawyers, teachers were also highly respected in the Black community (Drewry and Doerman, 2003). Therefore, if the motivation of those who inspired the students and those who governed the institution changed, it is only a natural progression that the HBCUs transformed into a vehicle that propelled students into the idea of a "superior" social status. Attending college for social gains was not exclusive to Black students; it was also reflected in the greater American community, especially after the introduction of the GI Bill in 1944. The GI Bill "encouraged public expectations and aspirations that they could attain the trappings of middle-class status" (Clark, 1998, p. 440).

Frazier (1957) suggests that a number of students of the Black elite at HBCUs were more concerned with being popular and in Greek organizations, than their college education and the disparity of America's Negro population (Frazier, 1957). Moreover, the former president of Philander Smith College, Lafayette Harris, suggested that some students at the Black college adopted an apathetic attitude toward the plight of the "submerged masses," as concerns of the Black community were often disregarded. Harris was quoted as saying, "with him [the Negro student], very little seems to matter except meals, sleep, and folly . . . They know nothing of their less fortunate fellowmen and care less" (Anderson, 1988, p. 212).

Despite the first Black colleges and universities being established during the mid-nineteenth century, the first Black Greek-letter organizations (BGLOs) did not appear at a Black college until 1907. Like other exclusive Black clubs, Greek organizations were another "way to distinguish themselves from nonmembers who could not afford the membership fees or pay for the kinds of clothes, parties, and automobiles that were *de rigueur* for members" (Graham, 1999, p. 86). They began as small social groups but were able to build "their popularity by seeking out certain desirable student candidates [that were] smart, popular, accomplished, affluent, athletic [and] good-looking, and turning down others" (Graham, 1999, p. 85). According to Little (1980), many dark-skinned Blacks and less financially able students, quite naturally, resented these discriminatory practices and founded rival clubs. Prior to 1940, BGLOs only intensified intraracial class and color divisions (Little, 1980).

Administrators at some Black colleges were apprehensive about allowing BGLOs on their campuses because of prior experience with elite social clubs (Little, 1980). For example, prior to 1925, "Fisk University students were prohibited from joining or belonging to any college fraternity or other secret college organization while at the University" (Little, 1980, p. 140). Alternatively, some HBCUs and various BGLOs based admission on color and class. According to Little (1980, p. 141), while fair-skinned Blacks were often given preference, "a black or brown-skinned co-ed had to come from a well-to-do family, have a better than average scholastic record, [and] be beautiful" (in rare cases, a person with a dark complexion could be considered beautiful if they had European-like features and long, flowing hair). In addition to BGLOs, some HBCUs founded in the nineteenth century based their offer of admission on one's complexion (Jones, 2000). Law professor Trina Jones notes that such institutions included Atlanta University, Fisk, Hampton, Howard, Morgan State University, Spelman, and Wilberforce University (Jones, 2000).

Although of concern, the Black elite's abandonment of other Blacks and what was considered Black culture seems to be a natural progression. During the first half of the twentieth century their parents placed a premium on middle-class values and New England traditions (Frazier, 1957). Moreover, their parents' parents tended to emphasize the ways of middle- to upper-class Whites because they were taught White values in the master's home (Gatewood, 2000). According to Johnnetta Cross Brazzell (1992), White middle-class values were taught to make Blacks more competitive in the greater American society. Black students prior to the 1920s attended colleges that "devoted virtually no attention to the cultural heritage of Africa, but emphasized Anglo-Saxon or American culture. The educational experience of the black upper class, then, conspired to mold it into a replica of middle- and upper-class white America" (Gatewood, 2000, p. 279). In his 1930 commencement speech at Howard University, Du Bois discussed how Blacks had lost sight of the goal of education as an instrument for racial uplift, because of indoctrination by White American values and selfish actions (Thelin, 2004).

Similar to the slavery era, during the early to mid-1900s European features were the epitome of the concept of beauty and glamour. With the exception of "exotic" Black women like Lena Horne and Dorothy Dandridge, Whites and mainstream America never equated beauty with the Black woman (Golden, 2004; Moss et al., 1975). For that reason, *Ebony* magazine set out to prove that Black women were beautiful, through the use of "photo studio tricks" and natural charm. In 1945, *Ebony* nominated a female college student from Sarah Lawrence College, named Barbara Gonzales, to take pictures with a top photographer. *Ebony* hailed the pictures as comparable to those in White fashion magazines. Despite the success of the photographs, Gonzales was not completely Black; she was also part Venezuelan, Indian, Chinese, and Spanish (*Ebony*, 1945). The media's depiction of Blacks and its gauge of attractiveness during Gonzales's era were drastically improved in comparison to the nineteenth and early twentieth century. During that time, the media, specifically magazines like *Harper's* or *Atlantic* portrayed Blacks "as bug-eyed, big-lipped and sometimes drawn as beasts with tails" (Golden, 2004, p. 41).

BLACK COLLEGES POST WORLD WAR II

The controversial scholar Frazier suggested that after World War II fair-skinned Blacks remained focused more on etiquette and proper speech, while the "common" Black student was more focused on uplift for themselves, their family, and community, by earning a college education. Nevertheless, the profile of the average Black student at Black colleges began to change in the 1940s and 1950s, as enrollment increased and the economic background of college students began to diversify (Freeman, 2005). Despite the 1954 *Brown* decision, which overturned the "separate but equal" decision of *Plessy v. Ferguson*, "most black college students continued to attend HBCUs years after the decision was rendered" (U.S. Department of Education, 1991). However, it was the GI Bill and perhaps the 1964 Civil Rights Act that afforded tens of thousands of Black students the opportunity to receive a college education and increased Black college enrollment (Freeman, 2005; "The rising number," 1997; U.S. Department of Education, 1991). The GI Bill, introduced in 1944, reshaped the role of higher education in America, by funding one year of college enrollment cost for every 90 days of service. After World War II, veterans of color and those of all socioeconomic status enrolled en masse, and therefore it reshaped the role of higher education in America, as college and the idea of college was made accessible to all (Clark, 1998; Thelin, 2004). The Civil Rights Act of 1964, legislation that ensured equal rights and opportunities for all, afforded Blacks an opportunity to attend historically White institutions (HWIs). However, it wasn't until the 1970s when Black enrollment was greater at HWIs than HBCUs (Freeman, 2005). So while the *Brown* decision and the Civil Rights Act of 1964 encouraged college enrollment at White colleges with superior facilities and larger budgets, perhaps it also planted the seed of higher education in general. And considering the tumultuous times, perhaps Blacks were more comfortable in what they thought to be a supportive Black college environment, hence the increase in Black college enrollment until the mid-1970s.

The 1960s bore the Black Power Movement, which called for pride in African and Afro-American traditions, unity within the Black community, and a new sense of racial consciousness. During this time, the popular adage, "Say it loud, I'm Black and I'm proud," developed as many African Americans displayed their sense of natural beauty and their rejection of White culture through wearing afros and dashikis (Keith and Herring, 1991). Some fair-skinned Blacks faced questions of authenticity and were made to prove their Blackness because of learned mistrust by their dark-skin kin (Jones 2000). Although the narrow perspective on issues concerning skin color and race positively changed for many dark-skinned Blacks and Blacks of lower socioeconomic backgrounds during this time, little changed for a minority of the Black elite. With the passing of the *Brown* decision in 1954, the extreme Black elites were given their one-way ticket to escape the "common" Blacks flaunting their African heritage, and therefore, they fled to HWIs. This is not to say that all Black students who attended HWIs were of an elitist frame of mind or looking to escape the "common" Blacks. For example, Golden, author of *Don't Play in the Sun: One Woman's Journey through the Color Complex*, suggested that she chose to attend American University, an HWI, over Howard because she wanted to circumvent the color conscious environment of Howard University (2004). However, the mass exodus of Black students from HBCUs to HWIs did not truly begin until the early 1970s (Freeman, 2005).

Although issues of complexion and class have never been and probably will never completely dissipate, the sense of Black pride during the Black Power movement afforded dark-skin and fair-skin Blacks the opportunity to come together in celebration of their race and ethnicity. According to Joel Rosenthal (1975), as Blacks have fought to alter their inferior status in American society, they have chosen models that have alternately emphasized and de-emphasized their group's ethnicity. Through the Black Power movement, dark-skinned Blacks became more assertive on matters concerning the advancement of the Black race, more specifically dark-skinned Blacks (Blackwell, 1985). According to Cheryl Mobley (personal communication, December 12, 2005), a member of Alpha Kappa Alpha sorority (AKA) and Howard University student during the late 1970s, for the first time in Howard University's history, a dark-skinned woman was crowned Miss Howard in 1977. Ms. Mobley states that the contest winner, a member of Delta Sigma Theta, was victorious because of the assertiveness of the dark-skinned students at the university. Upon receiving her crown, she was booed and jeered by the undergraduate audience and was met with extreme disappointment from Howard's administration, despite being very talented and well-liked at the university. It was customary that a fair-skinned student be crowned Miss Howard because that was the student body's impression of attractiveness (personal communication, December 12, 2005). Mrs. Mobley, who considers herself fair-skinned, spoke of the challenges she was met with at Howard because she was one of a few AKAs who were friends with members of Delta Sigma Theta. She described the Deltas as typically being of a darker hue.

ENTERING THE 1980S

In the 1980s, the color issues that plagued the Black community prior to the Black Power Movement resurfaced through popular culture. Research shows that Black "college students tend to judge attractiveness partly in terms of skin color, the most admired color being lighter than average but not at the extremely light end of the scale. These students also compromise between reality and wish-fulfillment and rate their own skin color in the direction of the preferred shade" (Freeman et al., 1966, p. 365). This is very evident in Spike Lee's movie *School Daze* (1988), where Greek organizations are used as a lens to address issues of color and class in the African American community. The two groups of women were the Wannabees or the Gamma Rays and the Jigaboos, modeled respectively after the sororities, AKA and Delta. As previously mentioned, AKA is known for its light-skinned members with long, straight hair. On the contrary, Delta is known for its medium- to dark-skinned members with short hair. According to Gregory Parks and Clarenda Phillips (2005, p. 417),

> The Gammites and Gamma Rays flaunt crass materialism, are politically (a)pathetic, are presented as a mimicry of white fraternal members, and spend the majority of their time engaging in unproductive hazing and pledging rituals. Lee calls the have-nots the Jigaboos. These dark-brown-hued college students are Afro centric and politically focused; the commit their energy and activities to demanding that their college, financially divest from South Africa.

According to a *Hilltop* (Howard University student newspaper) journalist, Nina Goodwine (2005), issues of color and class, and specifically the use of the brown paper bag test, existed at Howard in its early days, but have been eradicated in theory in the present day. Although Byron Stewart, the student body president, is quoted in Goodwine's article as saying that he does not believe color is an issue at Howard or in society, other Howard students and graduates are quoted as saying the opposite. Jennifer Jordan, an African American literature professor at Howard, believes little has changed regarding the paper bag theory (Goodwine, 2005).

Colorism and Classism in Today's Society

Although color and class elitist attitudes may not be as evident today at HBCUs, they are still prominent in today's society. For example, the accusations surfaced that *Vanity Fair* (2005) lightened pop singer Beyonce Knowles's (the first Black woman to appear on *Vanity Fair*'s cover in 12 years) skin color with the intention of increasing newsstand sales. According to *Radar* magazine, "the pop diva's medium-dark complexion was air-brushed to a 'Jennifer Lopez shade of bronze' to fit in with the magazine's cheery new aesthetic" (http://www.radaronline.com/fresh-intelligence/2005/10/14, retrieved November 3, 2005). To the contrary, darkening a Black person's skin has also been a ploy used to increase newsstand sales. Unlike *Vanity Fair*, the cover of the June 27, 1994, issue of *Time* magazine exaggerated O. J. Simpson's skin color to play into the "dangerous Black male" stereotype, to subconsciously persuade its audience that Simpson was guilty of murder. After a series of experiments, Keith B. Maddox, a professor of psychology at Tufts University, argued that "both Whites and light-skinned Blacks have deep-seated beliefs in stereotypes that dark-skinned blacks are more prone to violence, criminal activity, drug use, and laziness" (2002, p. 46).

Psychologists Kenneth and Mamie Clark found that when Black children were presented with both Black and White dolls of varying skin tones, the children distinctly preferred the White dolls (Jones, 2000). The Clarks found that Black children overwhelmingly associated the White doll with positive characteristics. In the 1980s, Michael Barnes duplicated the Clark study. "Barnes concluded that children, unlike adults, may be more honest about their racial self-hatred" (Russell et al., 1992). Many children of color are intensely attentive to the relationship between color and privilege, and want to benefit from the rewards connected to whiteness (Hill, 2002). This results in the promulgation of colorism, the questioning of one's self-worth, and feelings of being a disappointment for yet another generation of Black children. While this unfortunate hierarchy of skin color is pervasive in the Black community, it is not an issue restricted to Blacks. Complexion ranking is very evident in Brazil, South Africa, Italy, Japan, the Philippines, the Middle East, and the West Indies (Bates, 1994; Charles, 2003; Daniel, 2003; Leong, 2006; Texeira, 2003). In Brazil, race is not based upon ethnicity, country of origin, or lineage, but on education level and economic status. In Brazil, it ranges from Black to mulatto to White, with other classifications in between. Those of African heritage are believed to be socially inferior, hence those with little to no education and little income are considered Black and those with the most influence and high socioeconomic status are "blessed" as being White. The fluidity of race in Brazil allows Brazilians of the darkest complexion and kinky hair to "gain" the status of White, should they obtain educational credentials and establish themselves financially (Daniel, 2003; Texeira, 2003).

CONCLUSION

Through reviewing the livelihoods of African slaves and acknowledging their fortitude in persevering through hardships created by their White masters, one can begin to understand their logic in the fight for the right to receive an education and to achieve a higher socioeconomic status in the United States. Once excluded from attaining a higher education at historically White colleges and universities, upper- and middle-class Black students have created their own niche within the college setting and within the world by assimilating to the American middle-class values and ideals.

Constructions of colorism and classism have evolved, but the effects, low self-esteem, and missed opportunities remain the same. So while Black students are no longer barred from colleges based on color or complexion, undertones of colorism remain at colleges, Black and White, and in society in general. As evidenced through the lightening of pop star Knowles's magazine cover picture and the darkening of Simpson's magazine cover picture, the issues plaguing the "field slaves" and the "house slaves" are exclusive to our nation's Black colleges, but inclusive to all in today's global society.

Colorism, and to some extent classism, remain taboo and provocative topics among Blacks. However, in order to reach the end goal of a harmonious Black community, Blacks must no longer circumvent the issue and address it directly. To address the issue means to understand its complex existence and history, to question one's definition of beauty and self-worth, and to actively challenge erroneous beliefs. Only then can we attain equality within the Black community and the greater world. It was Du Bois who identified the color line as the problem of the twentieth century; and it is distressing to see such concerns continue into the twenty-first century (Du Bois, 1978). Therefore, while these issues have a long ugly history in America, it is up to today's generation to discontinue its perpetuation. Martin Luther King Jr. once famously said, "I have a dream that my four children will one day live in a nation where they will not be judged by the color of their skin but by the content of their character." America has come a long way since his famous "I have a dream speech" in 1963; however, we must not stop; we must continue this progression.

NOTES

1. When Virginia master Ralph Quarles's freed mulatto daughter married a slave, he purchased the slave's freedom and gave them their own plantation with slaves. R. Toplin, "Between Black and White: Attitudes toward Southern Mulattoes, 1830–1861," *Journal of Southern History* 45 (1979): 185–200.
2. South Africans were required to register as a member of one of the designated races, African, Coloured, Indian, or White. Such racial classifications were done until the mid-1990s. Classification varied depending on the locale and year; the criteria ranged from appearance, lineage, race of spouse, socioeconomic status, and "biology". The biological tests included examining one's nails, eyelid, and genital pigmentation.

L. Thompson, *A History of South Africa* (New Haven, CT: Yale University Press, 2000).

3. Women could not attend Harvard and Yale at this time; however, Oberlin pioneered the practice of coeducation. Yale and Harvard did not shift to the practice of coeducation until 1969 and 1972, respectively. J. Thelin, *A History of American Higher Education* (Baltimore, MD: Johns Hopkins University Press, 2004).

4. The "Seven Sisters" colleges are Barnard College, Bryn Mawr College, Mount Holyoke College, Radcliffe College, Smith College, Vassar College, and Wellesley College. J. Thelin, *A History of American Higher Education* (Baltimore, MD: Johns Hopkins University Press, 2004).

REFERENCES

Allen, W. 1992. The color of success: African American college student outcomes at predominantly White and historically Black colleges. *Harvard Educational Review* 6: 26–44.

Anderson, J. 1988. *The education of Blacks in the South.* Chapel Hill, NC: University of North Carolina Press.

Anonymous. 1946, July. Negro pulchritude. *Ebony*, 27.

Bates, K. 1994, September. The color thing. *Essence*, 25, 79.

Bennett, J. 2001. Toni Morrison and the burden of the passing narrative. *African American Review* 35(2): 205–17.

Blackwell, J. 1985. *The Black community: Diversity and unity.* New York: Harper & Row.

Charles, C. 2003. Skin bleaching, self-hate, and Black identity in Jamaica. *Journal of Black Studies* 33(6): 711–28.

Clark, D. E. 1998. "Two Joes meet–Joe college, Joe veteran": The GI Bill, college education, and postwar American culture. *History of Education Quarterly* 38(2): 165–89.

Cross Brazzell, J. 1992. Bricks without straw: Missionary-sponsored Black higher education in the post-emancipation era. *Journal of Higher Education* 63(1): 26–49.

Daniel, R. 2003. Multiracial identity in global perspective: The United States, Brazil, and South Africa. In *New faces in a changing America*, ed. L. I. Winters and H. L. DeBose, 247–86. Thousand Oaks, CA: Sage.

Definition of Colorism by the Center for Race and Gender. 2005. Retrieved October 24, 2005, from http://crg.berkeley.edu/programs/programs.html.

Drewry, H, and H. Doermann. 2003. *Stand and prosper: Private Black colleges and their students.* Princeton, NJ: Princeton University Press.

Du Bois, W. E. B. 1978. *W. E. B. Du Bois on sociology and the Black community.* Chicago: University of Chicago Press.

Du Bois, W., ed. 2003. *The Negro problem.* New York: Humanity Books. Original work published in 1903.

Du Bois, W. 2005. *The souls of Black folk.* New York: Pocket Books.

Editor. 1997. The rising number of African-American college diplomas. *Journal of Blacks in Higher Education*, 17, 45.

Frazier, F. E. 1957. *Black bourgeoisie: The book that brought the shock of self-revelation to middle-class Blacks in America.* New York: Free Press Paperbacks.

Freeman, H. E., D. Armor, J. M. Ross, and T. F. Pettigrew. 1966. Color gradation and attitudes among middle-income Negroes. *American Sociological Review* 31(3): 365–74.

Freeman, K., ed. 1998. *African American culture and heritage in higher education research and practice.* Westport, CT: Praeger.

Freeman, K. 2005. *African Americans and college choice: The influence of family and school.* Albany, NY: State University of New York Press.

Gasman, M. 2006. Salvaging "academic disaster areas": The Black college response to Christopher Jencks and David Riesman's 1967 Harvard Educational Review article. *Journal of Higher Education* 77(2): 317–52.

Gasman, M., and S. Anderson-Thompkins. 2003. *Fund raising from Black-college alumni: Successful strategies for supporting alma mater.* Washington, DC: Council for Advancement and Support of Education.

Gatewood, W. B. 2000. Aristocrats of color: The educated Black elite of the post-reconstruction era. *Journal of Blacks in Higher Education* 29(August 2000): 112–18.

———. 2000. *Aristocrats of color: The Black elite, 1880–1920.* Fayetteville, AR: University of Arkansas Press.

Golden, M. 2004. *Don't play in the sun: One woman's journey through the color complex.* New York: Anchor Books.

Goodwine, N. 2005. The legacy of the brown paper bag. *Hilltop,* September 16. Available at http://www.thehilltoponline.com/media/storage/paper590/news/2005/09/16/Campus/The-Legacy.Of.The.Brown.Paper.Bag-987550.shtml.

Graham, L. 2000. *Our kind of people: Inside America's Black upper class.* New York: Harper Collins.

Harvey, R. D., N. LaBeach, E. Pridgen, and T. M. Gocial. 2005. The intragroup stigmatization of skin tone among Black Americans. *Journal of Black Psychology* 31(3): 237–53.

Hickman, C. 1997. The devil and the one drop rule: Racial categories, African Americans, and the U.S. Census. *Michigan Law Review* 95(5): 1161–265.

Hill, M. 2002. Skin color and the perception of attractiveness among African Americans: Does gender make a difference? *Social Psychology Quarterly* 65(1): 77–91.

Hudson, B. 1963. Another view of "Uncle Tom." *Phylon* 24(1): 79–87.

Hunter, M. 2002. If you're light you're alright. *Gender and Society* 16(2): 175–93.

Johnson, W. B. 1996. *Black Savannah, 1788–1864.* Fayetteville, AR: University of Arkansas Press.

Jones, T. 2000. Shades of brown: The law of skin color. *Duke Law Journal* 49(6): 1487–557.

Keith, V., and C. Herring. 1991. Skin tone and stratification in the Black community. *American Journal of Sociology* 97(3): 760.

Kephart, W. 1948. Is the American Negro becoming lighter? An analysis of the sociological and biological trends. *American Sociological Review* 13(4): 437–43.

Kerr, A. 2005. The paper bag principle: Of the myth and the motion of colorism. *Journal of American Folklore* 118(469): 271–89.

Larsen, N. 1969. *Passing.* New Brunswick, NJ: Rutgers University Press.

Lee, S. (Director/Producer/Writer). 1988. *School Daze* [Motion picture]. United States: Columbia Pictures.

Leong, S. 2006. Who's the fairest of them all? Television ads for skin-whitening cosmetics in Hong Kong. *Asian Ethnicity* 7(2): 167–81.

Little, M. 1980. The extra-curricular activities of Black college students, 1868–1940. *Journal of Negro History* 65(3): 135–48.

Maddox, K. B. 2002. Brown paper bag syndrome: Dark-skinned Blacks are subject to greater discrimination. *Journal of Blacks in Higher Education,* 37, 46.

Moss, M., R. Miller, and R. Page. 1975. The effects of racial context on the perception of physical attractiveness. *Sociometry* 38(4): 525–35.

Parks, G., and C. Phillips. 2005. *African American fraternities and sororities: The legacy and the vision.* Lexington, KY: University Press of Kentucky.

The rising number of African-American college diplomas. 1997. *Journal of Blacks in Higher Education,* 17, 45.

Rosenthal, J. 1975. Southern Black student activism: Assimilation vs. nationalism. *Journal of Negro Education* 44(2): 113–29.

Russell, K., M. Wilson, and R. Hall. 1990. *The color complex.* New York: Double Day Publishing.

Sirk, D. (Director), and F. Hurst (Writer). 1959. *Imitation of Life* [Motion picture]. United States: Universal International.

Stuckey, S. 1987. *Slave culture: Nationalist theory and the foundations of Black America.* New York: Oxford University Press.

Texeira, M. T. 2003. The new multiculturalism: An affirmation of or an end to race as we know it? In *New faces in a changing America,* ed. L. I. Winteres and H. L. DeBose, 21–8. Thousand Oaks, CA: Sage.

Thelin, J. 2004. *A history of American higher education.* Baltimore, MD: Johns Hopkins University Press.

Toplin, R. 1979. Between Black and White: Attitudes toward Southern mulattoes, 1830–1861. *Journal of Southern History* 45(2): 185–200.

U.S. Department of Education, Office for Civil Rights. 1991. *Historically Black colleges and universities and higher education desegregation.* Retrieved May 6, 2007, from http://www.ed.gov/about/offices/list/ocr/docs/hq9511.html.

Wesley, C. 1940. The concept of Negro inferiority in American thought. *Journal of Negro History* 25(4): 540–60.

Williams, J. 2004. *I'll find a way or make one. A tribute to historically Black colleges and universities.* New York: Harper Collins.

Willie, C., and D. Cunnigen. 1981. Black students in higher education: A review of studies, 1965–1980. *Annual Review of Sociology* 7: 177–98.

A White wash at *Vanity Fair?* 2005. *Radar,* October 14. Retrieved November 3, 2005, from http://www.radaronline.com/fresh-intelligence/2005/10/14.

CHAPTER 12

NOT A LAUGHING MATTER: THE PORTRAYALS OF BLACK COLLEGES ON TELEVISION

ADAM PARROTT-SHEFFER

From *The Cosby Show* to Black Entertainment Television's (BET) *College Hill*, historically Black colleges and universities (HBCUs) have been part of popular culture. Popular representations of African Americans in the media have often relied on, as writer Sylvester Monroe (1994, p. 82) wrote in *Essence* magazine, the tropes of "clowns, cut-ups and loud mouthed buffoons." When the situation comedy arrived on television screens, this role tended to remain the same. An emphasis on slapstick characterizations left little room for more complete portrayals of African American identity. Examples are prevalent and the antics of lead characters on *Amos and Andy* or *Sanford and Son* consistently relied upon these characterizations.

Black colleges and universities, as organizations committed to the education of African Americans, have had relatively few representations on the small screen. Beginning with the fictional Hillman College on *The Cosby Show* and *A Different World*, the spin-off that takes place at the same college, to the modern "reality" situations of BET's widely controversial *College Hill*, there have only been three significant portrayals of HBCUs on television. Additionally, these portrayals tend to be inaccurate in their sensational representations of the people who attend HBCUs and in depictions of campus life. In this chapter, I attempt to place the television portrayals of HBCUs within the context of scholarship pertaining to Black colleges in order to assess whether HBCUs are presented fairly in this form of popular media. I argue that representations of HBCUs on television are not aligned with what the research on HBCUs suggests, and that much of this disparity stems from the tension between HBCUs being viewed as intellectual entities and television's traditional casting of African Americans as clowns. Even the lack of portrayals of HBCUs on television can be attributed to this tension.

It should be noted that there are several film portrayals of HBCUs, including *Drumline* and *School Daze*. For this chapter, I have chosen to focus only on television portrayals. Since *College Hill* appears on television, the show's inclusion in this study is more appropriate than any one of the movies on Black colleges even though it is not a situation comedy as are the other two shows discussed within the scope of this study.

I begin my study with a chronological overview of the three shows that have depicted HBCUs on television: *The Cosby Show, A Different World,* and *College Hill.* I then place these three representations within the historical context of their development. In other words, I attempt to analyze the shows relative to the specific point in time in which they aired. The purpose is to provide context to root any analysis of the representations by linking them to the environment that allowed for their creation. Next, I analyze the depictions of the HBCU environment as presented by these three television series. Specifically, I examine the rationale students used in choosing to attend an HBCU, as well as portrayals of academics and of student life. Finally, I argue that the simplistic renderings of HBCUs on television, whether positive as in the case of *The Cosby Show* or negative as on *College Hill,* have a negative effect on Black colleges and their very survival. This stereotyping limits or distorts the purported special purpose of these institutions to provide academic and economic uplift for a specific population by reducing these schools to mere tropes of college life. In this chapter, I break the shows down primarily by subject matter rather than chronologically; this method allows for an easier comparison of portrayals and contributes to a richer understanding of the complexity of media representations.

In order to gather a richer description of television portrayals of HBCUs, I will look beyond episodic evidence from the three programs that most directly portray Black colleges, incorporating news articles, press releases, published interviews with those affiliated with the programs, scholarly journal articles, and advertisements. My goal is to further flesh out how these particular programs contextualize HBCUs for their respective audiences. I examined how these sources portrayed HBCU students, these students' reasons for selecting an HBCU, and the academic rigor of these institutions.

HBCUs ON TELEVISION

The Cosby Show, debuting in 1984, is oftentimes heralded as a new era for African Americans on television. Numerous scholars, including Sut Jully and J. Lewis (1992) and Linda Fuller (1992), have explored the popular appeal of the show, its emphasis on middle class African American life through its portrayal of the Huxtables, a fictional New York family, and its ability to celebrate African American culture without ridicule within the confines of the situational comedy. The treatment of HBCUs on the program was no less revolutionary. HBCUs became an element of *The Cosby Show's* storyline on May 15, 1986, in an episode titled "Denise's Decision." In the course of applying to college, Denise considers matriculating at her parents' alma mater, the historically Black Hillman College. After weighing several school options, Denise decides to attend Hillman. The next few years of programming include several visits to the campus as well as prominently placed paraphernalia, including sweatshirts, pennants, and pictures.

A Different World was a spin-off of *The Cosby Show*, originally examining Denise's (played by actress Lisa Bonet) experiences at Hillman College (Gray, 2004). The show first aired in 1987, and ended in 1993. The show centered on Denise's room-mate Jaleesa (Dawnn Lewis) and dorm mate Whitley (Jasmine Guy) after Lisa Bonet left the show due to a pregnancy. Some scholars praised *A Different World* as the "ideal" Black sitcom. Writers such as Robin Means Coleman (1998, p. 104) com-mended the show for its "social relevancy, uncompromising construction of Blackness (Whiteness was not the normative yardstick), and the ability to use the situation comedy formula as a vehicle to highlight the Black experience in America." The HBCU setting allowed *A Different World* to provide a unique in-depth account of higher education.

The cancellation of *A Different World* marked the end of an HBCU presence on network television for the remainder of the 1990s. In fact, the next representation of HBCUs in television entertainment to appear is the highly controversial reality show, *College Hill*, which premiered on BET on January 28, 2004. BET billed *College Hill* as the first Black dramatic reality TV show. The show's setting at the campuses of Black colleges provides a rare glimpse into the television medium's portrayal of HBCUs. *College Hill* provides viewers with the "real" life stories of students attending HBCUs. How real the reality presented on the program actually is has been the source of much of the controversy surrounding the show; a topic further explored later in the chapter.

THE COSBY SHOW: MIXED REVIEWS

Much of the literature on *The Cosby Show*, which relates to the show's portrayal of HBCUs, praises its constant commitment to Black higher education, but also sim-plifies it to little more than name dropping. Leslie B. Inniss and J. Feagin (1995, pp. 695–6) explain that "through Cosby's wearing of collegiate sweatshirts . . . the show consistently sent out messages about the importance of Black academic institu-tions." The implication is that the Huxtables' success was due to their educational experiences at HBCUs. Yet, sweatshirts and road trips were not enough for some viewers. A minority thought of these trinkets as token gestures that did not portray Blackness appropriately and critiqued, "for some, *The Cosby Show* failed to adequately intersect being Black in America even while being a member of the Black middle class" (Coleman 1998, p. 103). In short, while *The Cosby Show* featured snippets of Black culture, these critics asserted that the show was not a true exploration of Black culture, but was predicated on Blacks acting White. One might question the validity of these reviewers' assertions.

The notion of "acting White" is oftentimes a statement on what is assumed to be "true" Black culture. "Acting White" is a derogatory way of referring to intelligent African Americans that pursue higher education. A 1986 study by Signithia Fordham and John Ogbu asserted that Black students oftentimes performed poorly in school because to be academically successful was tantamount to selling out. For *The Cosby Show*, this accusation of "acting White" implies that the show portrayed White plots and White characterizations despite the use of Black actors and writers. While the validity of the theory of "acting White" as a cause of low academic performance is well outside the scope of this chapter, as the charge relates to *The Cosby Show* the evidence is limited in supporting this claim.

What the previous research has not analyzed is how *The Cosby Show*'s portrayal of Hillman College challenges this view of Black culture. Two episodes of the sitcom directly feature HBCUs. The first one, Episode 49: "Denise's Decision," has already been mentioned. This episode chronicles Denise's agonizing over her decision of choosing a college. With the family's help, she finally decides upon Hillman as the place for her. The second, Episode 74: "Hillman," features the Huxtables visiting the college to attend a ceremony for the retiring president of the institution. Each family member shares their memories of their time on campus, stories from professors about Cliff Huxtable's (Bill Cosby) academic prowess and interest in Denise's experience at Hillman. The episode ends with the president's farewell speech regarding the mission of Hillman graduates.

A *DIFFERENT WORLD*: A BROAD EXAMINATION

As the spin-off of *The Cosby Show*, many of the same praises and critiques that applied to *The Cosby Show* apply to *A Different World*. The sitcom follows roommates Denise, Jaleesa, and Maggie (Marisa Tomei) through the trials and tribulations of their freshman year at Hillman College. They clash with wealthy Southern belle housemate Whitley (Jasmine Guy), their dorm director, and the unwanted advances of the loveable, but hopelessly nerdy Dwayne Wayne (Kadeem Hardison). The first season develops the friendship between the three roommates through clashes surrounding Denise's messiness, Jaleesa's age and marital status (26 and divorced), and Maggie's nonconfrontational personality. What never enters into their conflicts is race; Maggie is the only recurring White character on the show. The emphasis of the sitcom is on the unity of the students despite their mild disagreements. Hillman is portrayed as a supportive environment with numerous friends, maternal dorm directors, and paternal professors. Hillman is a school any parent would find acceptable for their child to attend. Given *A Different World*'s status as the number two hit of the 1987–88 season (with 23.4% on the Nielson ratings), there is little doubt that it was also a show college parents watched each week with their kids (Fuller, 1992).

Yet, *A Different World* has not garnered the same scholarly interest as *The Cosby Show*, appearing ancillary to discussions of *The Cosby Show*. Still, within the framework of network television's portrayals of HBCUs, *A Different World* takes center stage. Like *The Cosby Show*, *A Different World* features varied portrayals of HBCU life. Producer and Director Debbie Allen described the purpose of the show for E! Entertainment's *A Different World: I Was a Network Star*. She explained, "I did not want to see young Black students portrayed as people who were just drinking and partying and babysitting eggs. It had to be about something" (UNCLE Film + Television, 2005). While Allen did not take over until the second season of *A Different World* (the egg reference is to an episode in the first season),[1] she did have creative control for most of the show's duration. Under Allen's stewardship the sitcom addressed topics such as race, AIDS, and poverty. It did so all while reflecting the culture of an HBCU environment. Allen's portrayal of Black colleges features a foundation of African American culture, a strong student support network from a variety of sources on campus, and an embedded theme of the need to develop leaders.

COLLEGE HILL: THE "REAL" HBCU

From its initial January 28, 2004, premiere, *College Hill* has received press for its portrayal of HBCU campuses. The first season takes place at the 95% African American Southern University in Baton Rouge, Louisiana. The show transforms the eight students into tropes: the cheerleader, the jock, the rapper, the pregnant one, the wild child, the frat guy, the rich girl, and the nerd. While *College Hill* does not fall under the situation comedy category, like the other shows used for this study, its emphasis on "reality" situations and its setting at an HBCU warrants inclusion in this chapter. The series is also one of the few contemporary examples of HBCUs in popular culture. I would argue that reality shows are the sitcoms of the twenty-first century with their prime-time slots, basis in real-life situations, characterization, popular appeal, and common subject matter. An analysis of *College Hill* is also warranted because reality programming has moved into the primetime spots that were once the domain of sitcoms. Reality programs depend upon narratives that can be wrapped up with simple solutions within the span of an episode or two. These programs also rely on developing characterization through basic plot, character interaction, and fanciful situations.

Stephen Hill, senior vice president of Music Programming and Talent at BET, noted in an interview regarding *College Hill* that "there is college life, and then there's Black college life. . . . Those who have been around historically Black colleges and universities know there are social, cultural and attitudinal differences from the more mainstream institutions" (PR Newswire, 2004, p. 1). This recognition of differences between HBCUs and historically White institutions by those responsible for the production of the program implies that part of the goal of the show is to highlight those differences for the viewer; this intentional focus on the uniqueness of HBCUs makes the BET portrayal of Black colleges all the more significant. With *College Hill*, BET attempts to capture the experience of attending an HBCU.

Still, both viewers and critics alike must question how accurate the *College Hill* portrayal of HBCUs is. The situations real life participants are placed in on a reality program are oftentimes more fictionalized than the events that occur to fictional characters in the average situational comedy. For example, the second episode of *College Hill* features male students riding in a limousine with alcohol and Budweiser girls. *College Hill's* focus on the sensational and "sexy" continues throughout most of the season. In fact, it continues into the second season at a new campus. Year two of *College Hill* involves moving the setting of the show from dorms to an off-campus ranch at Langston University in Oklahoma. The cast of season two also features its own set of trope characters including the flirt, the "baller," the "rumpshaka," the cheerleader, the single mother, the pretty boy, the freak, and the good girl. The characterization of the cast provides a cursory glimpse of the values the media associates with HBCUs and those who attend them. What emerges seems to be a simplistic, one-dimensional, and sensational rendering of Black colleges. I will further critique this image by examining how these three shows deal with specific aspects of HBCU life: the reasons students attend, the academics, and student life.

REASONS FOR ATTENDING

Most research suggests that many students who choose to attend an HBCU do so because of geography, religious background, and the school's social reputation, independent of gender, income, and educational aspiration (McDonough et al., 1997; Freeman, 2002). Patricia M. McDonough (1997) also asserts that HBCU students are more likely to have a college-educated father and come from the Southeast or Southwest. In "Black Colleges and College Choice: Characteristics of Students Who Choose HBCUs," Kassie Freeman (2002) argues that knowing someone who attends an HBCU, a strong drive toward seeking roots, or a lack of cultural awareness are predictors of HBCU attendance. The television portrayals of HBCUs examined in this chapter both resonate with these findings and oversimplify them.

The Cosby Show identifies several reasons why Hillman becomes the school of choice for several of the family members. First, while never explicitly stated, legacy seems to be an implied part of the characterization of the Cosbys. Cliff's father Russell Huxtable, Cliff, his wife Clair Huxtable, and their daughter Denise all attend the school for some duration of fictional time. In the episode "Hillman," Rudy, the youngest daughter, explains to the whole family that someday she will attend Hillman as well (Sandrich, 1987). The show's emphasis on the members of the Cosby family attending HBCUs does seem to fall in line with the research regarding HBCUs and actual rationales for students choosing to attend a historically Black college. The portrayal of legacy at Hillman embedded throughout the program seems to resonate with Freeman's findings regarding the correlation between attendance and knowing someone else who attended an HBCU (2002). The geographic research findings of McDonough seem to be reflected in the cast of *College Hill.* Of the 16 cast members to appear on the first two seasons of *College Hill,* eight were from the Southwest, five from the South, two from the Midwest, and only one was from the Northeast. Within television programming, the overall impression of students' reasons for attending an HBCU seems to be aligned with what scholars suggest are the actual reasons for attending an HBCU. Where television and reality begin to differ is what the students do once they arrive on campus.

ACADEMICS

The Cosby Show attempted to portray HBCUs with the same high degree of positive images that it afforded to other aspects of Black culture including speech, art, music, and authors. When the Huxtables visit the campus of Hillman in the third season finale, the positive role of the HBCU in their fictional family's success is a recurring theme. Cliff reminisces to Theo, his son (Malcolm Jamal Warner), that one of his favorite professors constantly told his students that "the brain is the most important muscle in your body and I am going to whip it into shape" (Sandrich, 1987, episode 74). The emphasis on the rigor of academic life at HBCUs continues in the episode when the same professor Cliff remembers fondly and reveals Cliff's not-so-academic pursuits as an undergrad, but evokes that Cliff's success in life is proof "that a Hillman education can save anybody" (*The Cosby Show,* episode 74). The theme of high academic standards for students remains a focal point through the end of the program, when the new president of the college reminds those gathered at the retirement ceremony of her predecessor that "an A student at Hillman is an A student anywhere" (*The Cosby Show,* episode 74).

These were the words of wisdom she received from the former president that gave her the confidence to continue with graduate school. The finale of season three leaves no doubt in the viewer's mind that HBCUs should be considered places of high standards and great academic achievement. The episode "Hillman" portrays HBCUs as places of academic merit and enhanced student support so positively that few advocates of these organizations could find fault with the portrayals. Yet, this representation is limited by the medium of the sitcom.

The campus scenes used to move from scene to scene in *The Cosby Show* and *A Different World* tend to portray students as similarly clean-cut, well dressed, and walking purposefully to an imagined class with backpack in tow. The implication is that there is a certain "look" to an educated person, or at least a student educated at an HBCU. Many of the images in these passing shots are reminiscent of the United Negro College Fund (UNCF) images of Black students during the 1940s and 1950s (Gasman and Epstein, 2004). The overall effect is that while these shows portray much of academic life at HBCUs positively, they rely on a stereotyped image of Black intellectualism. It is problematic because this portrayal—which is very similar to what is oftentimes presented by HBCUs themselves—potentially excludes students who do not fit this middle- to upper-middle class image. A quick glance at the websites of Hampton, Howard, or Spelman supports this point; there seems to be an accepted "look" by which students are judged.

The professors at Hillman are well connected to their students. In the second episode of *A Different World*, Denise makes a checkbook error and is unable to pay both her debt and Jaleesa's debt. This conflict is the focal point of the episode as Denise attempts to prove her independence and adulthood by solving the problem on her own. Unwilling to go to her father for help, Denise turns to a dean who also teaches at Hillman. The dean provides Denise with fatherly financial advice; his comments suggest that he knows Denise and her family rather well. The dean even puts her bill on hold and sets her up with a campus cafeteria job. The whole relationship is portrayed much more like a surrogate father than that of professor. In another episode of *A Different World* titled "Wild Child," a professor, who is also a dean, takes great effort in cultivating personal relationships with his students. He knows about them academically and socially. He even helps a homeless student enroll officially at Hillman so that she has a place to live. The mentoring function of professors portrayed in *A Different World* seems to mirror the research on actual benefits of attending an HBCU (*Journal of Blacks in Higher Education*, anonymous, 1994). As one of the more unique aspects of the HBCU experience, the representation of these student-teacher relationships on *A Different World* reflects accurately this element of HBCU culture. Critics of HBCUs could argue that these close relationships are related to college attendance at any small liberal arts school. However, it remains a significant aspect of life at Black colleges, one that is captured again and again in *A Different World*.

Academic life, as represented in *College Hill*, is also defined by the majors selected by the students chosen for the show. The first season includes two marketing majors, a computer science major, biology major (headed toward premed), a communications major, an elementary education major, and one undecided student. There is a clear emphasis on business and science-related fields and a complete lack of liberal arts disciplines. The liberal arts remain unrepresented in the second season as well. Year two has three education majors, three biology majors, a radiology major, and one business major. This emphasis on the health and business professions tends to portray a collegiate experience that is not about developing scholars and academics but on providing middle-class professionals with practical skills that have direct job applications. The program seems to support Gregory Price's (1998) critique of HBCU's transformation from the cultivators of Black intellectuals to the producers of the intellectually devoid whose only concern is material progress. Granted, the skills being taught are for doctors, business people, or other high-paying professions, but they are nonetheless content- and skills-based majors.

STUDENT LIFE

There is a wholesomeness to college life developed within the world of the Huxtables. In the episode titled "Hillman," Cliff and Clair reminisce about meeting each other back in their undergraduate days. In part, at least for the scriptwriters of *The Cosby Show*, the HBCU provides a good environment to find a successful, intelligent marital partner. Arguably many people meet their spouses in college.

Some would argue that the religious background of most private HBCUs is one element that makes the institutions unique in the experience they offer their students versus many traditionally White colleges and universities that do not have any religious affiliation. However, the writers' vision of HBCU life dramatically differs from the popular perception of college life when the writers choose to incorporate religious themes on such a grand scale. At the end of the episode, as part of a final send-off for the retiring president, Clair leads the Hillman choir in the singing of a spiritual. Given this section of the program was allotted approximately one-sixth of the show's 30-minute time slot, its relevance cannot be ignored. This link between campus life and religion at a Black college, while perhaps overly emphasized in the scene, is probably a useful cultural differentiator to include. Many HBCUs have strong foundations as religious institutions and have preserved close ties with Black churches.

With an emphasis on the student life at an HBCU, *A Different World* provides the viewer with numerous depictions of student interactions at a Black college. During the television series, there is a recurring theme of students helping one another at Hillman. In episode 4, Dwayne Wayne gets into trouble with the dorm director for staying past curfew in the girls' dorm in order to help Denise with a calculus exam. In episode 18, Maggie and Denise do their best to help Jaleesa get over her stage fright so that she can deliver a stellar speech for class. Even Whitley is willing to assist her antagonist Denise when Rudy comes for a visit and needs a place to stay despite the fact that much of the show's banter is built upon the disdain Denise and Whitley have for each other. This theme of helping one another transcends friendships on the show and is part of the ambience that makes up Hillman.

There is also a sense of conservatism portrayed by student life at Hillman. Students worry about curfew and phone calls to parents for poor performance or behavior. The girls debate whether men should be allowed on the upper floors of the dorm and if the 11 p.m. curfew for male visitors is wrong. While most schools had abandoned the idea of *in loco parentis* by the 1980s, fictional Hillman held on to them. In the show,

HBCUs are portrayed as a place of family values and wholesome entertainment. The worst thing the students ever do is steal the head of their rival's mascot, and even this event becomes a family affair as Grandpa Huxtable gets involved. This view of HBCUs seems more conservative than the reality on actual campuses in the 1980s and 1990s. In another vein, Black colleges portrayed in television series are not the racially diverse places they tend to be in practice. According to Michael Nettles and Laura Perna (1997), one-fifth of those attending HBCUs in 1994 were not African American. However, the diversity of Black colleges that Nettles alludes to was not featured on the television series. *The Cosby Show* hinted at the notion that non-Blacks attended HBCUs with crowd scenes featuring the occasional White student during the Hillman episode. *A Different World* cast one White (Italian) lead, only to drop her the following season. All of the students featured in the first two seasons of *College Hill* are Black. The first consequence of this trend is the defining of the color line as simply Black and White. Furthermore, the paucity of conversations concerning the theme of race can be attributed to the fact that these programs lack the racially diverse cast to be able to facilitate these conversations. When the theme of race is addressed on these programs, it is only within a Black/White dichotomy; this limits the complexity of the discussion by leaving out the impact of the enrollment of Latinos, Asian Americans, and Native populations in HBCUs. Yet even within this absence of discussion of race, a few characterizations of Whites attending HBCUs can be made.

Episode 14 of *A Different World*, "Wild Child," features Cougar Barnett, a street-smart Caucasian girl with exceptional academic ability whom Denise befriends in math class. Despite performing well in class, Cougar disappears after the professor, and dean, begins looking into her registration. Much to Denise's surprise, they find out that Cougar is not only unregistered for the class, but that she does not attend Hillman. In fact, Cougar is homeless. The episode ends with Denise and the dorm director finding a way to get Cougar into Hillman on scholarship. The one White guest appearance on *A Different World* portrays a woman of low economic status relying on the assistance of educated Blacks. This is a powerful reversal of racial roles as demonstrated on television and probably the first of its kind.

While Cougar's characterization on *A Different World* provided a unique perspective on race, the recurring characterization of Maggie, the sole White cast member, was somewhat more limited to typical racial roles. Maggie is a transfer student who is sometimes ditzy and sometimes brilliant. In an early episode she publicly defends a woman's right to a career, much to the chagrin of her almost fiancé. Maggie only tends to have problems or conflict within the story arch when Denise or Jaleesa are at fault. For example, in episode 21, Jaleesa loses Maggie's practice baby (an egg) much to the emotional distress of the "mother." Whenever there is a problem, Maggie is there to lay out the choices and summarize the arguments for both sides. The implication is that she provides answers to the dorm conflicts and is the voice of reason for the roommates. It is a typical "White" role for television.

For the creator of *College Hill*, part of what makes student life at an HBCU so dynamic is the make up of the student body of primarily first-generation college students. Tracey Edmonds (PR Newswire, 2005, p. 1), the director of *College Hill*, refers to students attending Black colleges as "dynamic, colorful, interesting, uninhibited, and articulate." Edmonds goes on to explain the appeal of the show and how it relates to the struggles these students face in earning college degrees at an HBCU. She argues that HBCU students are "less guarded" about their experiences and what they deal with in their lives (2005, p. 2). In short, because of their first-generation college student and lower economic status, HBCU students are more honest. While it is difficult to substantiate the veracity of that claim, it is possible to discuss the implications of that claim for popular perceptions of HBCUs. The power of this belief over popular conceptions is decidedly greater when the director of one of the few shows on television to address HBCUs as subject matter holds the belief that HBCU students are more honest than their White peers.

IMPLICATIONS FOR HBCUS

Most would agree that the entrance of HBCUs onto the television screen not only reflects a change in perspectives regarding the role of Black colleges in American society, but also profoundly influences how HBCUs are perceived by the general population. What has yet to be determined is whether this new perception, as influenced by the portrayals of HBCUs in the three programs included in this study, has increased the survivability of HBCUs or whether these representations lead to a loss of credibility. If there is a loss of credibility, it must be asked whether this loss ultimately prevents HBCUs from carrying out their original purpose of educating while keeping in mind the specific educational, social, and emotional needs of a specific population.

After season two of *College Hill* completed its run at Langston University, the alumni association expressed rather grave concerns about the depiction of their alma mater. One alumnus claimed that it was a "disgrace at best to put Langston in the light they have with nudity, ghetto language, and unrestricted use of alcohol" (Murphy 2005, p. 2). Alumni were concerned that this portrayal of their alma mater would cheapen the image of the school and the value of its pedigree. The BET Network addressed the concerns of the Langston alumni by including promotional advertisements of Langston.

Raymond Downs, the vice chancellor for student affairs at Southern University, the setting of the original *College Hill*, explained to the Associated Press that the alumni at Southern had raised the same image concerns as the alumni at Langston (2005). Downs also made it clear that enrollment increased since the program aired and that he asserted that this was due "in part because of the exposure the university received from the program" (Murphy, 2005, p. 3). What is not known is whether these newly matriculated students would have attended other HBCUs, a historically White college or university, or if they would not have considered college altogether. In other words, does this media exposure expand the outreach of HBCUs to underserved populations who would have otherwise not contemplated or accessed college life or is it simply reallocating students who would have received the benefits of higher education even without exposure to the show? What is clear is that the depiction of Black colleges on television has been a mixed blessing for the institutions. While a spot on the small screen has created a larger recruiting pool of students and has legitimized their existence through media exposure, this publicity has come with a price. HBCUs have lost some power over how they are portrayed and, therefore, how the schools are conceived of by the majority of the population. Yet, at least for Downs, what is important is "that the young people love it" and that "it was great in terms of enrollment" (Murphy 2005, p. 3). While HBCUs have always been at the mercy of the media through newspaper and magazine articles as Gasman's (2006) study of the role of media coverage in the fiscal problems at Morris Brown College exemplifies, the college's portrayal by television shows even further complicates the perception of HBCUs by the American public.

Allen made similar claims about how television increased enrollment at HBCUs in a letter accompanying the season one box set of *A Different World*. She asserts that "after six years on the air, we tripled the enrollment of historically Black Colleges and Universities" (Allen, 2005). While her figure contradicts the 23% increase in enrollment at HBCUs over the 1984–94 period documented by the Patterson Research Institute, she is correct in arguing that enrollment at HBCUs did go up while the show was on the air (Nettles and Perna, 1997, p. 59). What cannot be determined is how much of this increase can be attributed to increased exposure of HBCUs on television programming. Despite this uncertainty, it is likely that *A Different World* did aid the increase in enrollment at HBCUs in some fashion through mere exposure. Television portrayals of HBCUs have had an impact on more than college enrollment figures. There are additional implications of how these institutions are represented on the small screen.

For instance, *A Different World* has received some criticism on elements it shares in common with *The Cosby Show*. One PhD candidate in educational media questioned the color range within the Huxtable family, focusing much of her critique on Denise, the second eldest daughter. The candidate wondered "what message is sent to Black America that Denise was the character for whom *A Different World* was created" (Fuller, 1992, p. 127). Denise is the fairest skinned of the Huxtable children and this academic is concerned that there is a reinforcement of the false notion of the superiority of light skin tones on both *The Cosby Show* and *A Different World*. The assumption is that most Americans would not watch a show about a darker-skinned Huxtable attending an HBCU. Perhaps a show with a darker-skinned lead attending a predominately White institution could succeed. Unfortunately, the implication is that a darker-skinned lead in the Black college and university system would not capture viewers.

The depiction of HBCUs by *A Different World* become more complex as changes in direction took place over the six-year run of the program. While the first seasons were grounded in the wholesome innocence of Denise, Whitley, and Maggie, eventually Denise and Maggie moved on. Whitley married Dwayne Wayne and the show began to focus on a new set of more street-wise freshmen with Lena (Jada Pinkett) from inner-city Baltimore taking the lead. Portrayals of Black college life were more complicated in the later years of the show, as *A Different World* featured episodes about sexuality and condoms and guest appearances by Tupac Shakur. The early years of the sitcom focused primarily on the community developed on the Hillman college campus; the later years shifted focus to the issues surrounding those attending HBCUs: class, race, poverty, and sexuality. Of significance, the move allowed the show to switch from themes about how HBCUs affect students to how HBCUs can shape society. Both contexts create space for addressing the worth of HBCUs within American higher education.

Despite its relatively positive and rich depictions of HBCU life, *The Cosby Show* does offer varied representations of HBCUs. One of the more enlightening revelations regarding the purported role of HBCUs takes place in the "Hillman" episode during the speech of the retiring president. He positions the college within a framework of social justice, racial harmony, and peace on earth while claiming not to use his speech to discuss those issues in detail (Sandrich, 1987). Where the role of HBCUs gets confounded is in the president's speech to the alumni of Hillman. The president ends his talk by advising alumni to "pick up" recent Hillman graduates and to share whatever they can with the graduates. The vision of *The Cosby Show* is one where graduates support one another through a shared sense of camaraderie.

Where these representations of HBCU life are portrayed matters greatly. *The Cosby Show*, as the number one rated show in America for five years, was watched by people from various ethnic and social backgrounds. Because it was a spin-off of such a popular show and also appeared on NBC, *A Different World* had a rather wide audience as well. However, *College Hill*, as a cable television show on a network catering to an African American demographic, has had a more limited audience. Who receives the information (or watches the show) can be just as important as what the show depicts. I would argue that *College Hill* would have had a stronger negative impact on HBCUs if it had been shown on NBC instead of BET because its portrayal would have been accessed by a larger and more diverse audience. These viewers, with less exposure to African American culture than BET's viewers, may have had more trouble looking past the kind of negative stereotypes depicted on the show. Such pejorative portrayals might have confirmed viewers' preexisting biases.

Still, the overarching issue of Black colleges and television is not how the institutions are represented, but how little they are represented. Yes, negative characterizations do an incredible amount of harm to the character and prestige of an organization, but HBCUs must be a part of the national cultural dialogue before their representation can be changed. Access to increased airtime is the first step to improvement in how the media portrays HBCUs.

Colleges and universities thrive on popular perceptions and reputation. To not be included in national popular culture as represented by situation comedies, dramas, news, and sporting events means atrophy for these institutions. Perhaps the adage "any publicity is good publicity" does not hold true, but the importance of inclusion in the public space (the realm in which people and cultures interact with implications pertaining to power dynamics) and the ability to participate within the mainstream dialogue is of utmost value to all organizations. As important generators of a unique aspect of American culture, HBCUs must become a part of the national dialogue. Greater representation in the media, both positive and negative, is necessary for the continued survival of HBCUs in the twenty-first century.

LIMITATIONS OF THE MEDIUM OF TELEVISION

In his defense of the lack of realism on *The Cosby Show*, Alvin Poussaint, the doctor who advised for the program, reminds readers that "critical social disorders, like racism, violence, and drug abuse, rarely lend themselves to comic treatment; trying to deal with it on a sitcom could trivialize issues that deserve serious, thoughtful treatment" (Poussaint,

1988, p. 74). As any viewer of shows such as *Saved by the Bell* or *Full House* that are directed at teenagers can attest, sitcoms commonly use overly dramatic episodes to address complex social issues and find solutions within the plot of one episode. They are usually not an appropriate forum for complex societal issues such as race, homelessness, or drug use because the very nature of the genre is to simplify all things into a 30-minute plot. Even though this simplicity makes any scholarly examination of television's renderings of Black colleges more challenging, it does not negate the importance of portrayals of HBCUs on these programs. Television treatment may be limited in complexity, but HBCUs are powerfully defined on television simply by how their campuses appear, how the character of the student body is portrayed, and how the link between college life and intellectual pursuits is defined.

While it is easy to point out the limited and overly simplistic nature of representations of HBCUs on television, it is important to use television's portrayal of other less mainstream organizations as a litmus test for how different these representations of HBCUs actually are as compared to representations of other organizations representing a minority population. For example, women's colleges are also mostly nonexistent on the television screen. In general, portrayals of minority populations have been consistently lacking in the media. Mainstream television tends not to include people, persons, or settings to which the majority of viewers cannot relate.

I would argue that television programming, and specifically the situational comedy and reality television, tends to simplify characters, plots, and settings into easily identifiable tropes. The sitcom is a device that grants the consumer significant amounts of information in a decidedly limited time span. As their purpose is to entertain with humor, these programs naturally tend to focus on the comedic elements of all things. Finally, the short plot duration of television shows does not lend itself to in-depth examinations of subjects. A similar argument can be made regarding reality programming, except in reality programming the purpose is to entertain both with extreme drama and with the exposure of personal and awkward situations. These realities of both of these genres do not free these genres from critique. Instead, the discussion must focus on which images of HBCUs are simplified and incorporated into situational comedy plot devices and what is indirectly implied about HBCUs with subtleties of dress, style, scenery, and action.

CONCLUSION

HBCUs must become a part of the national dialogue through inclusion in American pop culture. Historically, Black colleges have had little presence in primetime television lineups. Only three programs have dealt with HBCUs with any depth of coverage: *The Cosby Show*, *A Different World*, and *College Hill*. Additionally, two of these shows were created by the same producers and the other only reaches a limited cable audience. Yet, even within these programs the portrayals of HBCUs have been relatively simplistic and one-dimensional. Whether that portrayal is as a wholesome place with a legacy of family learning as portrayed on *The Cosby Show* or the nonstop party of sex and alcohol featured on *College Hill*, the lack of portrayals of Black colleges in television series limits the public's perception of them as the diverse and unique institutions that remain of great value in contemporary society.

Throughout this chapter I have argued that the realities of HBCUs and the representations of Black colleges on television are not always aligned. Academic life is often portrayed as sterile and clean-cut at Hilman, or as nonexistent on the campuses portrayed on *College Hill*. The student populations also tend to be more monochromatic than they are in real life. The implications of television portrayals of HBCUs are twofold. On the one hand, Black colleges suffer from a lack of pop culture exposure. When HBCUs do receive media attention, it tends to have the positive effect of increasing the enrollment at these institutions. Yet, this media attention often leads to a loss of academic prestige in the popular perception of these institutions. Negative exposure on television can do immense harm to these institutions even compared with negative portrayals or claims from scholars or in newspapers. This is due to the broader exposure television has over these other mediums. Simply put, more people watch popular programming than read a given newspaper article or even glance at a scholarly journal. Newspapers and scholars also aren't afforded the additional exposure that reruns can provide.

Given the power of television, it is important that this medium continues to simulate reality, and that conscious decisions are made about how institutions serving minorities are portrayed. In other words, producers, directors, and networks must remain conscious of both whom the show is intended to reach and whom it actually reaches. The few portrayals of Black colleges on television remain adequate at best and racist at their most extreme. HBCUs will not benefit fully from media exposure until the representations of HBCUs in popular culture are as diverse and rich as the institutions themselves.

NOTE

1. The "egg episode" takes places chronologically within the first couple of episodes of season one, but was not aired until well toward the end of season one. The premise of the episode was that Maggie was taking care of an egg for one of her classes in an effort to teach her responsibility. She has her roommates babysit for her and the egg goes missing. The episode had very little to do with the collegiate life and even less to do with the HBCU experience.

REFERENCES

Allen, D. (Producer and Director). 2005. *A Different World: Season one big laughs on campus* [DVD Insert]. Thousand Oaks, CA: Urban Works Entertainment.

Anderson, J. 1988. *The Education of Blacks in the South, 1860–1935*. Chapel Hill, NC: University of North Carolina Press.

BET delivers dose of "real" to reality TV genre with College Hill, television's first-ever Black dramatic series. 2004. PR Newswire.

BET goes back to class with the return of College Hill, television's first-ever Black dramatic reality series. 2005. PR Newswire.

The Blacker the college, the sweeter the knowledge? 1994. *Journal of Blacks in Higher Education* 5: 49.

The Carsey Warner Company (Producer). 1987–93. *A Different World* [DVD].

Coleman, R. M. 1998. *African American viewers and the Black situation comedy.* New York: Garland Publishing.

E! Entertainment, UNCLE Films (Producer). 2005. *A different world: I was a network star* [DVD].

Fordham, S., and J. Ogbu. 1986. Black students' school success: Coping with the "burden of 'acting white'". *Urban Review* 18: 176–206.

Freeman, K. 2002. Black colleges and college choice: Characteristics of students who choose HBCUs. *Review of Higher Education* 25: 349–58.

Fuller, L. 1992. *The Cosby Show: Audiences, impact, and implications.* Westport, CT: Greenwood Press.

Gasman, M., and E. Epstein. 2004. Creating an image for Black higher education: A visual examination of the United Negro College Fund's publicity, 1944–1960. *Educational Foundations* 18: 41–61.

Gasman, M. 2006. Truth, generalizations, and stigmas: An analysis of the media's coverage of Morris Brown college and Black colleges overall. *Review of Black Political Economy,* 111–35.

Gray, H. 2004. *Watching race: Television and the struggle for Blackness.* Minneapolis, MN: University of Minnesota Press.

Inniss, L., and J. Feagin. 1995. The Cosby Show: The view from the Black middle class. *Journal of Black Studies* 25: 692–711.

Jully, S., and J. Lewis. 1992. *Enlightened racism: The Cosby show, Audiences, and the myth of the American Dream.* Boulder, CL: Westview Press.

McDonough, P., A. L. Antonio, and J. W. Trent. 1997. Black students, Black colleges. *Journal for a Just and Caring Education* 3: 9–36.

Monroe, S. 1994, March. Hollywood: The dark side. *Essence,* 82–4, 127–8.

Murphy, S. 2005. New reality show prompts outcry from Langston alumni. Associated Press.

Nettles, M., and L. Perna. 1997. *The African American education data book. Vol. 1: Higher and adult education.* Fairfax, VA: Frederick D. Patterson Research Institute.

Poussaint, A. 1988, October. The Huxtables: Fact or fantasy? *Ebony,* 72–4.

Price, G. 1998. Black colleges and universities: The road to philista? *Negro Educational Review* 29 (1/2): 9–21.

Sandrich, J. 1987. Hillman. *The Cosby Show,* episode 74.

CHAPTER 13

Sexual Behavior Patterns and Sexual Risk-taking among Women and Men at a Historically Black University

Nelwyn B. Moore, J. Kenneth Davidson Sr., and Robert Davis

Enrollment in college engages unmarried students in a timeless conundrum. Armed with personal values formed from life experiences, such as family, peers, and religion, students arrive on campuses that offer opportunities to increase their independence, while concurrently, strong peer pressure encourages sexual behaviors that possibly conflict with their values. Additionally, research now reveals that individuals engage in their most extensive identity exploration during emerging adulthood, the usual college age, rather than early adolescence as previously believed (Arnett, 2000). This convergence has been reflected in decades of increases in premarital sexual involvement of college students and their risk-taking sexual practices, such as nonprotected sexual activity and mixing alcohol and sexual intercourse (Cooper, 2002; Langer, Warheit, and McDonald, 2001; Reinsch et al., 1992; Robinson and Jedlicka, 1982).

As do college students in general, Black students struggle with a cultural duality when the culturally prescribed behavior patterns and belief systems in their home-based environment collide with the new knowledge, attitudes, and beliefs found on college campuses. As a result, they can experience split loyalties between the Black community and the world of academia (Bazargan et al., 2000). However, there is not a homogeneous culture of either Black or White Americans on university campuses, but rather, diverse subcultures that reflect various socioeconomic milieus, values, attitudes, and behaviors.

Thus, Black college students can experience numerous challenges in their transition from late adolescence to adulthood while surrounded by new social, political, and cultural milieus found on university campuses.

Research indicates that premarital sexual intercourse among adolescents varies according to race/ethnicity, with the highest level of activity among male adolescents being, in order, Blacks, Hispanics, and non-Hispanic Whites; and among female adolescents, Blacks, non-Hispanic Whites, and Hispanics, respectively (Mosher et al., 2005). And, Davidson and Moore (2004) found that among never-married college students, Black women and men were more likely to have had sexual intercourse than White women and men.

Weinberg and Williams (1988) suggested that the more liberal Black sexual patterns are associated with a distinct subculture, and that Blacks are more liberal than Whites because they are more detached from family controls. They attributed the larger female–male differences in sexual activity for Blacks than Whites to a social milieu that is more sexually permissive for Black men and a ready availability of sexual opportunities for them. Willie (1991) argues, however, against using a purely historical perspective to study Black families, claiming that differentiation in lifestyles among Blacks can only be understood by examining the macro-environmental settings within which they live and the groups within which they interact.

Samuels (1997) argues that the nature of the sexuality of Blacks, and their sexual values in particular, has often been misrepresented under the guise of science. For example, earlier writers, such as DeRachewitz (1964) and Jacobus (1937), claimed that the virility of Black men is greater than that of White men; Black women respond instantly and enthusiastically to all sexual advances; and Black men are obsessed with the idea of having sexual activity with a White woman. Samuels (1997) postulated that such perceptions represent philosophical positions rather than empirical evidence, and that much of the sexuality research about Blacks today represents a recycling of cultural beliefs regarding their sexuality, couched in scientific terminology interpreted through the veil of White middle-class values.

Epidemiologic data from the Centers for Disease Control substantiate that Blacks in the United States are disproportionately diagnosed with HIV and other sexually transmitted infections (STIs). For example, in 2000, despite comprising only 12% of the general population, Blacks accounted for 76% of gonorrhea cases, 72% of syphilis cases, 52% of HIV cases, and 38% of AIDS cases (Robinson et al., 2005). And, Black college students may also be disproportionately affected. In 2000, two cases of acute HIV infection were identified among North Carolina non-Hispanic Black college students, and by 2003 that number had increased to 49 cases, or 88% of the total reported cases (Centers for Disease Control, 2004). With the shift in the HIV/AIDS population in the United States from the White, gay male community to women and Blacks, research needs to be refocused to increase the understanding of risk-taking sexual behaviors of these populations (Valentine, 2003). On the basis of this need and the dearth of findings and misrepresentations about Black sexuality in the research literature, this study sought to determine the sexual behavior patterns and sexual risk-taking among Black college students. Because of limited enrollment of Black students at most colleges and universities, in order to obtain a large sample, a historic Black College and University (HBCU) was selected as the study site.

SEXUAL BEHAVIOR

In examining the few studies from the last 25 years that have focused exclusively on the sexual behavior of Black female and male college students, noticeable inconsistent differences are found in their sexual behavior, as reported in Table 13.1. It could be argued that this lack of a pattern of behavior may be attributable to methodological issues such as sample size, age of respondents, and site of the data collection, including type of college/university and its geographical location. While several studies of Black college students have recently been published (Fennell, 1997; Harvey et al., 1996; Soet et al., 1999; Bazargan et al., 2000), variables important to this current study, such as percentage of sexual intercourse and age at first intercourse, were either not reported or not reported by sex. Also, Price and Miller (1984) included both never-married and married respondents in their analyses, making their data unusable for comparisons with never-married Black college students.

Black college women were less likely to have experienced sexual intercourse than Black college men, and to have begun their sexual debut at a later age in the decades of the 1970s, 1980s, and 1990s (Table 13.1). The mean number of lifetime sexual partners reported in the 1980s for Black college women ranged from 4.8 (Belcastro, 1985) to 6.3 (Price and Miller, 1984) in comparison to 7.4 (Belcastro, 1985) to 11.0 partners (Price and Miller, 1984) for Black college men. More recently, 56% of Black college women and 70% of Black college men reported having had two or more sex partners during the past year (DiIorio et al., 1998). Data from the National Survey of Family Growth (NSFG) indicate that for Black women and men (ages 20–24, without regard to educational level), the median number of lifetime sex partners was 2.8 and 3.8 partners, respectively. But, only 22.2% of these women and 32.5% of these men had had two or more sex partners during the past year (Mosher et al., 2005). Belcastro (1985) reported the mean frequency of sexual intercourse per month was 8.5 times for Black college women and 7.5 times for Black college men, with 85% of women having had an orgasm during sexual intercourse. And, 18% of Black college

Table 13.1 Previous studies of sexual behavior of Black college students

Author/year	Percent/sexual intercourse		Mean age/first intercourse	
	Women	Men	Women	Men
Christensen and Johnson (1978)	79.2	91.7	N/A	N/A
Price and Miller (1984)	N/A	N/A	17.6	14.5
Belcastro (1985)	90.0	94.7	16.2	13.6
Ford and Goode (1994)	70.8	84.9	N/A	N/A
Johnson et al. (1994)	76.0	83.0	N/A	N/A
Lewis et al. (2000)	89.0	N/A	16.0	N/A
Braithwaite and Thomas (2001)	71.4	N/A	N/A	N/A

Note: N/A = Not available.

women reported sexual dissatisfaction as contrasted with 24% of Black college men (Price and Miller, 1984).

SEXUAL RISK-TAKING

Sexual risk-taking is reflected in the relationship with one's sex partner. Of Black college students, 36% of women and 60% of men indicated their most recent sex partner was uncommitted, as in a date, friend, or acquaintance (DiIorio et al., 1998). Taylor et al. (1997) found that Black college women were more likely than Black college men to ask potential sexual partners about their sexual histories, 75% and 54%, respectively.

In other sexual risk-taking among Black college students, Ford and Goode (1994) found that 64% of women and 60% of men had had sexual intercourse after consuming alcoholic beverages, with 48% of the women and 47% of the men indicating that they "got high." And 37% of Black college women indicated "sometimes" or "always" being intoxicated during sexual intercourse (Lewis et al., 2000).

With the escalation of AIDS as a social issue, the use of condoms to reduce sexual risk-taking became more of a research focus. The efficacy of consistent condom use has been substantiated in research with HIV-discordant couples, while inconsistent condom use has been found to offer insufficient protection (Robinson et al., 2005). Yet, the percentage of Black college students who indicated they "always" used a condom ranged from 24% (Lewis et al., 2000) to 26% (Johnson et al., 1994) for women and was 35% for men (Johnson et al., 1994). However, of Black college students, 50% of women and 63% of men indicated using condoms during their most recent sexual experience (Harvey et al., 1996) and 58% of women reported carrying a condom if they thought that sexual activity might occur (Taylor et al., 1997). In a study of condom use for dual prevention (pregnancy and STIs) among Black college students, 86% were dual-prevention users. Those who used condoms only to prevent pregnancy reported fewer sex partners and perceived fewer barriers to condom use (Whaley and Winfield, 2003). The major barriers to the use of condoms reported by Black college women were "negative views of condoms," "feel invincible," and "trust in partner," while Black college men indicated "trust in partner," "living for the moment," and "negative views of condoms" (Duncan et al., 2002).

Sexual risk-taking is reflected in the incidences of STIs, unintended pregnancy, and abortion. The percentage of Black college women diagnosed with an STI ranged from 17% (Johnson et al., 1994) to 38% (Lewis et al., 2000), compared with 45% of Black college men (Johnson et al., 1994). Black college men are more likely than Black college women to have received treatment for a sexually transmitted disease (Braithwaite et al., 1998). In a study of Southern Black college students in the early 1990s, 3.2% had been diagnosed as HIV positive (Johnson et al., 1994) contrasted with 2% for other comparable student populations (Valentine, 2003).

Among Black college women, 41% reported having had an unintended pregnancy, with 62% of these women indicating at least one delivery (Lewis et al., 2000). Although comparable abortion data for Black college students is not available in the literature for the 1990s or the 2000s, one study of Black college women from the mid-1980s found that 10% had had an abortion (Belcastro, 1985).

DIFFERENTIAL ORIENTATION TOWARD SEXUALITY

DeLamater (1987) claims that social institutions such as religion and family affect sexual attitudes and behavior, not by dictating a specific level of permissiveness but rather by creating and maintaining different orientations. For example, the degree to which college students are subject to the liberal cultural milieu of contemporary society is likely to be mediated by personal religiosity. Although Robinson and Calhoun (1983) found that the degree of religiosity had no significant effect on the sexual liberality of Black female or male college students, Laumann et al. (1994) discovered from a general population that Black women were almost twice as likely as White women to indicate that religion shaped their sexual behavior. Davidson, Moore, and Ullstrup (2004) found that White college women who defined themselves as more religiously devout and/or who were more involved in institutionalized religion were less likely to be sexually permissive. Religious attitudes also increase levels of sexual guilt (Daniluk, 1993), which is both an emotional and behavioral reaction that encourages some sexual behaviors and discourages others, that is, nonuse of contraceptives (Davidson et al., 2004). Individuals experiencing high levels of sexual guilt are not likely to repeat behavior that produced the guilt (Cado and Leitenberg, 1990).

THEORETICAL FRAMEWORK

Although reference group theory (Zaleski and Schiaffino, 2000) and cognitive development theory (Sandler et al., 1992) are two frameworks that have been used to study sexual behavior, Davidson's (1993) middle-range cultural milieu theory regarding participation of women in premarital sexual intercourse was selected for this study. According to this theory, White female college students are more likely to experience premarital sexual intercourse when their development prompts independence from parents at the time that they are more subject to the liberal standards of their peers. Thus, sexual behavior is influenced by two interrelated factors: the developmental trajectory and social forces. While the applicability of the Cultural Milieu Theory has not been established in research on sexual behavior of White men or of Black women and men, its theoretical underpinnings appear to offer similar explanatory value for these other groups (Davidson, 1993). This study attempted to determine the validity of such assumptions. Specifically, this investigation sought to clarify the sexual behavior patterns of Black college women and men, and the differences, if any, in circumstances of first sexual intercourse, risk-taking sexual practices, and physiological and psychological sexual satisfaction.

METHODOLOGY

PROCEDURES

An anonymous questionnaire was administered during regular university classes to volunteer respondents enrolled in courses in introductory sociology, general psychology, economics, and health education. Some closed-form questions were formatted as individual Likert scale variables, with answer categories such as "strongly agree" to "strongly disagree" or "always" to "never." However, none of these variables represent components of any type of scale, but rather single variables in the survey instrument. The questionnaire included open-form and closed-form questions in the following areas: demographics, sexual history, contraceptive practices, sexual guilt, STI history, pregnancy history, and sexual satisfaction.

Sample

The initial sample for this investigation consisted of 312 female and 255 male respondents at an historical Black land-grant state university, which was founded in 1891. With its reputation as an outstanding PhD-granting institution and with a residential campus located in an urban setting of 223,000 people, the university attracts thousands of applicants annually. This popularity has led to a state law mandating that at least 82% of its entering freshman class must be from North Carolina. At the time of the data collection, there were 6,889 undergraduate students enrolled, 53% women and 47% men. Those respondents whose race/ethnicity was other than Black (6 women and 13 men) or whose marital status was other than never-married (30 women and 15 men) were declared as missing values in the data analyses. Further, to make the data set compatible with the existing research on college student sexuality, a more homogeneous sample was created, by declaring as missing values those who were over age 23 years (45 women and 25 men), whose class standing was special or graduate (1 woman and 4 men), and/or whose sexual orientation was either lesbian (4 women), bisexual (5 women and 1 man), or gay (1 man). The declaration of these missing values produced a subsample that comprised 255 female and 194 male Black respondents between the ages of 18 and 23 years. The percentage of the subsample that reported having had voluntary consensual sexual intercourse was 90.2% ($N = 229$) for women and 89.7% ($N = 174$) for men. Because of the focus of this study, the data analyses were limited to respondents who had experienced sexual intercourse.

The class standing of the female respondents ($N = 229$) in the final subsample included Freshmen 19.7% ($N = 45$), Sophomores 27.9% ($N = 64$), Juniors 25.8% ($N = 59$), and Seniors 26.6% ($N = 61$), while the male respondents ($N = 174$) included Freshmen 36.2% (N = 63), Sophomores 25.9% ($N = 45$), Juniors 24.1% ($N = 42$), and Seniors 13.8% ($N = 24$) [X^2 (3, $N = 403$) = 18.346, $p < 0.000$]. The mean age of the respondents was 20.2 for women and 19.8 for men (F (1, 401) = 9.726, $p < 0.002$).

Statistical Analyses

Given the non-randomness of the sample and the levels of measurement, the chi-square test was chosen to ascertain the significance of any differences for nominal variables while analysis of variance (ANOVA) was utilized for ordinal and continuous interval variables. Further, multivariate analysis of variance (MANOVA) was used to test for interaction effects of select variables. It should be noted that p values of 0.000 have been reported in those instances where the p value is zero to three decimal places using 0.5 or greater as the basis for rounding values upward.

FINDINGS

RELIGIOUS BACKGROUND

The most often reported religious denominations that the respondents were reared in were, in order, Baptist, Fundamentalist [e.g., Church of Christ and Nazarine],

Table 13.2   Religious background by sex

Variable	Women %	Women N	Men %	Men N	Group totals %	Group totals N	Group differences X²	d.f.	P
Religious denomination reared in							10.810	6	.094
Mainline Protestant	12.2	26	12.2	19	12.2	45			
Baptist	55.4	118	46.2	72	51.5	190			
Catholic	4.2	9	4.5	7	4.3	16			
Institutional Sect	7.0	15	5.1	8	6.2	23			
Fundamentalist	15.0	32	16.0	25	15.4	57			
Non-Christian	.5	1	1.9	3	1.1	4			
None	5.6	12	14.1	22	9.2	34			
Religious compared to others							11.185	2	.004*
More Religious	13.3	29	13.8	23	13.5	52			
About as Religious	71.1	155	56.9	95	64.9	250			
Less Religious	15.6	34	29.3	49	21.6	83			
Current level/religious commitment							6.294	2	.043*
Devout	17.7	36	11.0	17	10.1	53			
Moderately Devout	66.5	135	64.3	99	65.5	234			
Not Devout	15.8	32	24.7	38	19.6	70			

*$p < .05$

Mainline Protestant [e.g., Episcopalian, United Methodist, and Presbyterian, USA], and Institutional Sect [e.g., Church of God and Full Gospel] (Table 13.2). Women attended religious services more frequently than men, $M = 40.0$ times per year compared to $M = 29.9$ times per year [$F(1, 313) = 9.125$, $p < 0.003$]. Men were more likely than women to consider themselves to be less religious than others as well as less religiously devout (Table 13.2).

FIRST SEXUAL INTERCOURSE

An examination of the circumstances surrounding the first sexual intercourse revealed a number of significant differences between women and men. Women were older and had an older sexual partner at first intercourse than men (Table 13.3). Women also were more likely to have been in a committed love or steady dating relationship with their first sexual partner. Few women or men were under the influence of alcohol or drugs, and women were more likely than men to report that contraception was used, either by themselves or by their partners. Interaction effects were found between the variables, "First Intercourse/Used Contraception" and "Religious Compared to Others," with

Table 13.3   Circumstances surrounding first sexual intercourse by sex

Variables	Women %	Women N	Men %	Men N	Group totals %	Group totals N	Group differences X^2	d.f.	P
Relationship to partner at time							74.894	5	.000
Committed love	34.4	76	10.2	17	24.0	93			
Steady dating	34.4	76	21.1	35	28.7	111			
Occasional dating	12.7	28	10.8	18	11.9	46			
Friend	10.4	23	38.6	64	22.5	87			
Casual acquaintance	4.5	10	12.7	21	8.0	31			
Person just met	3.6	8	6.6	11	4.9	19			
First S.I. voluntary							3.756	1	.053
Yes-implied consent	51.4	108	61.5	99	55.8	207			
Yes-verbal consent	48.6	102	38.5	62	44.2	164			
Under influence/alcohol/drug							1.328	1	.249
Yes	5.3	12	8.2	14	6.5	26			
No	94.7	215	91.8	157	93.5	372			
Used contraception							7.156	1	.007*
Yes	72.4	163	59.6	102	66.9	265			
No	27.6	62	40.4	69	33.1	131			

Variables	Women M	Men M	Group mean	Group differences F	d.f.	P
Age at time	15.9	15.2	15.6	15.296	1, 376	.000*
Age of first sex partner	17.8	15.9	17.0	64.377	1, 380	.000*
Experience physiologically satisfying	2.5	3.8	3.0	101.482	1, 393	.000*
Experience psychologically satisfying	2.7	3.6	3.1	47.589	1, 392	.000*
Less anxiety during experience	2.9	3.2	3.0	6.650	1, 392	.010*
Less guilt during experience	3.3	4.3	3.8	57.118	1, 396	.000*

*p < .05

respondents who considered themselves less religious than others being more likely to have used contraception at their first intercourse experience [F (2, 378) = 5.995, $p <$ 0.007]. And, although not significant, the findings strongly suggest that women were more likely than men to have given verbal consent for their first sexual intercourse experience (Table 13.3). Women experienced significantly more anxiety and guilt than men and also reported their sexual experience to be both less physiologically and psychologically satisfying (Table 13.3). Interaction effects were also found between the variables, "First Intercourse/Psychologically Satisfying" and "Religious Compared to Others," with those respondents who considered themselves less religious than others being more likely to report their first intercourse to be physiologically satisfying [F (8, 377) = 9.861, $p <$ 0.011].

SEXUAL-RISK TAKING BEHAVIORS

The mean monthly frequency of sexual intercourse was 8.7 times for women and 10.8 times for men [F (1, 351) = 2.779, $p <$ 0.096]. Significant differences were found regarding the mean number of both lifetime and past-year sexual partners, with women having fewer sexual partners in both cases (Table 13.4). In addition, women were less likely to have had sexual intercourse with someone they had just met and more likely to have discussed their number of lifetime partners with their most recent sexual partner. Interaction effects were identified between the variables, "Engage in Sexual Intercourse/Person Just Met" and "Guilt During First Intercourse," with those reporting less guilt during first intercourse being more likely to have engaged in intercourse with person just met [F (4, 392) = 5.413, $p <$ 0.000]. Further, interaction effects were found between the variables, "Told Most Recent Sex Partner/Number/Lifetime Partners" and "Guilt During First Intercourse," with respondents who reported less guilt during first intercourse being more likely to discuss their number of lifetime partners [F (10, 386) = 6.433, $p <$ 0.019]. Women more often than men asked a new sexual partner about their number of lifetime sexual partners, with 52.6% of women "always/almost always" asking, compared with only 37.0% of men. In addition, the findings strongly suggest that women were more likely than men to inquire about the STI history of a new sexual partner (Table 13.4). However, 17.3% of women and 25.0% of men "never" inquired about STI histories, whereas 51.6% of women and 45.0% of men "always/almost always" asked. Contraception was no more likely to have been used in their most recent sexual intercourse by either women or men (Table 13.4). Of these women, 77.2% revealed that they had "sometimes" provided a condom for a sexual partner.

With regard to alcohol consumption, women drank less often than men, F = 26.8 times and M = 60.4 times per year, respectively [F (1, 304) = 40.339, $p <$ 0.000], and they were much less likely to become "moderately intoxicated" than were men [F (1, 310) = 55.177, $p <$ 0.000]. However, women were much more likely than men to indicate having been diagnosed with an STI (Table 13.4). The most often reported STIs by women were, in order, chlamydia, candidiasis, trichomoniasis, and gonorrhea, and by men, gonorrhea and genital warts. There were no cases of HIV or AIDS reported by women or men. While the differences were not significant between women and men with regard to either having ever been pregnant or having gotten a

Table 13.4 Risk-related sexual behavior by sex

Variable	Women %	Women N	Men %	Men N	Group totals %	Group totals N	Group differences X²	Group differences d.f.	Group differences P
Contraceptive used/most recent S.I.							.336	1	.562
Yes	69.3	149	66.5	103	68.1	252			
No	30.7	66	33.5	52	31.9	118			
Ever diagnosed w/STI							12.992	1	.000*
Yes	25.8	58	7.8	6	21.2	64			
No	74.2	167	92.2	71	78.8	238			
Ever been pregnant/gotten partner pregnant							.534	1	.465
Yes	32.3	64	36.3	45	33.9	109			
No	67.7	134	63.7	79	66.1	213			
S.I./person just met							51.039	1	.000*
Yes	29.6	67	65.7	109	44.9	176			
No	70.4	159	34.3	57	55.1	216			
Told recent partner number/lifetime sex partners							13.739	3	.003
Yes, Fewer	11.6	26	11.8	19	11.7	45			
Yes, Actual	54.2	122	40.4	65	48.4	187			
Yes, Greater			2.5	4	1.0	4			
No	34.2	77	45.3	73	38.9	150			

	Women	Men		Group differences		
	M	M	Group mean	F	d.f.	p
Number of lifetime sex partners	5.6	7.5	6.4	15.123	1,342	.000*
Number of different sex partners in past year	2.1	3.5	2.7	40.084	1,339	.000*
Frequency of asking new partner/ number of sex partners	3.6	3.1	3.3	10.897	1,385	.001*
Frequency of asking new partner about STI history	3.4	3.1	3.3	3.148	1,293	.077

*$p < .05$

sexual partner pregnant, the percentage of respondents, in general, with a history of unintended pregnancy was considerable (Table 13.4). Of these pregnancy outcomes, 68.3% of those for women and 65.6% for men resulted in induced abortions, and 14.6% of the pregnancies for women and 14.8% for men were carried to term, resulting in single motherhood.

SEXUAL SATISFACTION

Of the women, 49.3% reported "almost always" experiencing orgasm during sexual intercourse, while 23.6% indicated "sometimes," 12.9% "rarely," and 14.2% "never." Using a Likert scale variable ("never comfortable" to "always comfortable"), no significant difference was found between women and men concerning their level of comfort with their sexuality [$F(1, 398) = 0.001$, n.s.]. Further, using Likert scale variables to also assess their overall levels of current physiological and psychological sexual satisfaction ("very satisfied" to "very dissatisfied"), no significant differences were found [$F(1, 336) = 1.788$, n.s.; $F(1, 330) = 0.098$, n.s.].

DISCUSSION

Because of the paucity and incomparability of data in the literature, it is problematic to compare some of the findings of this investigation with studies of the sexual behavior of Black college students in former decades. However, as in the decades of the 1970s–1990s, Black college women in this study were less likely to have experienced sexual intercourse than Black college men, and to have begun their sexual debut at a later age. The percentage of sexual intercourse reported by these women (90.2%) approximates that found by Belcastro (1985) in the mid-1980s and was almost identical to that of Lewis et al. (2000) in the year 2000. For men, the reported percentage of 89.7% was somewhat below the 94.7% found by Belcastro (1985). With no later comparable data, it is difficult to speculate about this differential. However, the significantly younger mean age for men than women in this sample plausibly explains this puzzling discrepancy. A perusal of Table 13.1 will reveal that a decreasing spiral in mean ages for first intercourse (Belcastro, 1985; Lewis et al., 2000) was continued with the mean ages of female respondents in this study (15.9 years), but not for men (15.2 years). In fact, in these findings, the past discrepancy in age at first sexual intercourse between women and men has disappeared, confirming the claims of Upchurch et al. (1998). That sexual behavior among Black college women has increasingly become more like that of Black college men parallels the same convergence found among White female and male college students (Langer et al., 2001).

First intercourse experience appeared to differ in several ways for women and men. Not only did women wait until later ages to have sexual intercourse, as noted by Robinson and Calhoun (1983) in the 1980s, when they did so it was more likely to be with a committed partner or a steady dating partner. First sexual intercourse was also "safer" for women than men, with fewer women under the influence of alcohol and more of them reporting the use of a contraceptive. But why did women in this investigation feel more guilt and experience less physiological and psychological sexual satisfaction than men with their first sexual intercourse experience? Socialization factors surrounding the sexual double standard, as suggested by Johnson and Staples (2005), are suspected but not specifically addressed in this study. Certainly these findings about guilt cannot be observed without recalling the facts reported about religiosity. While the majority (70.4%) of both women and men in this investigation were from religious denominations considered to be sexually conservative, women more than men rated themselves more religious and devout in their current level of religious commitment than their partners.

While religion and guilt are not related in the research literature on Black college student sexuality, the two have been positively correlated among college students in other racial/ethnic groups (Davidson et al., 2004). Interaction effects revealed that women and men who felt less guilt during their first intercourse experience were more likely to discuss number of previous sex partners with new partner, a safer-sex behavior, but were also more likely to have intercourse with a person just met, a high-risk behavior. Such findings raise an arguable query: Is the emotion of guilt a positive or negative influence in terms of risk-related sexual behaviors? Interaction effects also revealed that respondents who were less religious were more likely to use a contraceptive; begging the question, how are religion and unplanned sexual intercourse related? Guilt is suspected but not proven in this study.

Positive movement in the direction of responsible sexuality was noted by these college women who reported fewer risk-related sexual behaviors than men, that is, being less likely to have had sexual intercourse with a person just met and having had fewer sex partners, both lifetime and in the past year, confirming the findings of Belcastro (1985). That women had significantly fewer lifetime sex partners than men suggests that perhaps men were more socially scripted in terms of permissive sexual behavior, supporting the findings of Sprecher and Hatfield's (1996) thesis that the sexual double standard is still a reality. The findings also underscore Hyde's (2005) assertion that gender differences are strikingly large for attitudes about having sexual intercourse in a casual, uncommitted relationship. Although an alarming number of women and men never inquired about the STI histories of their new sex partners, women did so more often than men. When with new sex partners, the women were more likely to both reveal and seek information concerning the number of lifetime partners, substantiating similar findings by Taylor et al. (1997).

While earlier research on Black college students revealed that 50% of women and 63% of men used a condom at their most recent sexual experience (Harvey et al., 1996), it seems promising for sexual health that larger percentages of women (66.5%) did so in this study. Although a range of 17% (Johnson et al., 1994) to 38% (Lewis et al., 2000) for STIs for Black college women has been reported, this study found that 26% of women had been so diagnosed, a significantly greater percentage than for men. Recent data for Black college men were not found with which to compare these findings. Others have reported that among Black college students, men were more likely to have received treatment for an STI (Braithwaite et al., 1998). Why were women in this study significantly more likely to have been diagnosed with an STI? While speculative, four scenarios are plausible. First, sexually active college women are more likely than men to use campus or local family planning clinic services. And, clinics offering oral contraceptives typically require a pelvic examination, at which time STIs potentially can be detected. Second, sexually active men are less likely to seek medical services because some STIs are more likely to be asymptomatic in men. Third, men were possibly less honest in their self-reporting. And, fourth, a time-related factor may have been the most likely single cause. Women in this sample, 18% freshmen and 28% seniors, were both older and advanced in class standing than men, 36% freshmen and 14% seniors.

Finally, the percentage of female and male participants who had been involved in an unintentional pregnancy did not differ significantly, but the rate for women (32.3%) was considerably lower than the 41% reported by Lewis et al. in 2000, a positive change. Unfortunately, comparable 2000 or later data were not available for men. But the earlier findings of Belcastro (1985) that only 10% of Black college women chose to terminate their unintended pregnancy with an abortion is in sharp contrast to the 68.3% of women in this study. That two-thirds of women and men involved in an unplanned pregnancy indicated that it had resulted in an abortion is striking. The wide disparity in abortion figures from yesterday and today poses questions. Does the higher abortion rate found in this sample simply reflect the greater societal acceptance of abortion among women in general or among Black women specifically in the years since the 1973 *Roe v. Wade* decision? If the latter, does the 58% increase since the 1980s substantiate the claim of Upchurch et al. (1998) that the racial gap in sexual behavior is disappearing? This and other similar questions can lend direction to future efforts in education and research about family planning.

CONCLUSION

In concluding that fewer Black college women than men are sexual risk-takers, one must not overlook maturity as a possible intervening variable. The disparity in age and class standing, with significantly older women than men in this sample, may have been a determining factor in the practice of safer sex. But, in addition to age, these findings suggest that religiosity may have fostered the safer sexual behaviors for women on every dimension than for men. As postulated by Davidson's (1993) Cultural Milieu Theory, perhaps for women a combination of their developmental trajectory (older age for women) and their social milieu (religious environment) prompted their safer sexual behavior patterns. The question then becomes, were the women merely more successful than the men in negotiating what Bazargan et al. (2000) characterized as the "cultural duality," differences in their home-based environment and their college-campus environment? If so, why? Because of these findings, this query deserves further study.

Future research needs to focus more directly on family background factors and the interface of religion, asking questions that would clarify the ways in which family systems and religious institutions function that possibly result in different orientations toward sexuality, and different outcomes in the sexual behavior of Black college students. University personnel in charge of freshmen orientation could use these data to support the use of more direct parental involvement in their programs to help students negotiate the cultural duality they face when leaving home to enter college. Hopefully, religious leaders in schools of theology and in local churches will also find these results to be of interest in their programming efforts to incorporate a healthy sexuality model as an integral part of their religious education curriculum and programs.

In a 2003 study, using a sample of Black students from 11 North Carolina colleges, the three most common reasons identified for continuing to engage in high-risk sexual behavior were "lack of sustained prevention messages targeting young Blacks," "feeling personally disconnected from the reality that they might contract HIV," and "believing that physical characteristics and appearance can reveal sex partner's HIV status" (Centers for Disease Control, 2004, p. 731). The third reason reported in the North Carolina study calls into question the commonly assumed knowledge-behavior gap whereby information is known but not translated into safer-sex behaviors. Although unknown, it is possible that the higher level of risk in sexual behavior of men than women in this study was partially related to the men's lack of knowledge since they were significantly the youngest respondents. It is, however, also possible that their behavior was motivated by the "magical thinking" of invincibility characteristic of younger adolescents, verifying Arnett's (2000) data that emerging adulthood is the time for completing developmental tasks formerly assigned to early adolescence. To understand the significance of such cognitive and emotional developmental issues in sexuality education, additional research is needed.

If, in fact, lack of sustained prevention messages and feeling personally disconnected are key issues in the continuing high-risk sexual behaviors of Black college students, student affairs practitioners are challenged to do "outside of the box" creative thinking about campus programming. For example, modeled after the highly successful "Writing across the Curriculum" program on college campuses, a pilot "Sexuality across the Curriculum" program could offer the needed sustained prevention messages and help young Blacks feel more connected to reality. As a major campus emphasis involving appropriate faculty/staff, such programming could help reframe sexuality within contexts whereby a variety of sexuality issues could be explored from different perspectives. If successful, this curriculum, functioning as an intervention program, could be shared with other Black colleges.

Finally, armed with empirical data, student affairs practitioners and other university personnel will be positioned for roles of leadership in community efforts to find common ground from opposing sides on the issue of sexuality education in public schools, an essential element in a democracy that values the physical and psychological sexual health of all of its citizens. Whether at lower or higher educational levels, Olson (2005) suggests that a first step in accomplishing the goal of health-based sexuality education is to acknowledge that sexuality education begins in philosophy, not in curricula or research. He contends that sexuality must be recognized as a central feature of the human experience and reframed as contextual, relational, familial, social, emotional, and spiritual. Hopefully, this study has made a small contribution to the concerted efforts required of many groups such as researchers, educators, practitioners, and clergy to assure healthier sexual behavior of young Black college students.

As some have argued, because only 12% of Black college students are enrolled at HBCUs, sexuality data collected from these sites cannot be generalized to the majority of Black college students (Kenya et al., 2003). Neither can they be applied to those in other geographical locations without reservation. But, the relatively large size of this Black sample compared with most others in the research literature lends additional weight to these findings. And, while this investigation focused on the sexual risk-taking practices of today's Black students at a Black college, the results have as much and perhaps more relevance to other Black college student populations than the plethora of White college student research in the literature. While many answers remain elusive, a number of questions have been raised for future research and practice. The real significance of this study, however, may not be measured in statistical terms. Its greatest contribution may be to create an awareness of a major neglected area in sexuality research: the study of today's Black college students, leaders of tomorrow.

REFERENCES

Arnett, J. J. 2000. Emerging adulthood: A theory of development from the late teens through the twenties. *American Psychologist* 55: 469–80.

Bazargan, M., E. Kelly, J. Stein, B. Husaini, and S. H. Bazargan. 2000. Correlates of HIV risk-taking behaviors among African-American college students: The effect of HIV knowledge, motivation, and behavioral skills. *Journal of National Medical Association* 92: 391–404.

Belcastro, P. A. 1985. Sexual differences between Black and White students. *Journal of Sex Research* 21: 56–67.

Braithwaite, K., and V. G. Thomas. 2001. HIV/AIDS knowledge, attitudes, and risk-behaviors among African-American and Caribbean college women. *International Journal for the Advancement of Counseling* 23: 115–29.

Braithwaite, R., L., T. T. Stephens, K. Braithwaite, and V. G. Thomas. 1998. Behavioral predictors of intended sexual behavior among a sample of African American college undergraduates. *Journal of Black Psychology* 24: 164–78.

Cado, S., and H. Leitenberg. 1990. Guilt reactions to sexual fantasies during intercourse. *Archives of Sexual Behavior* 19: 49–63.

Centers for Disease Control. 2004. HIV transmissions among Black college student and non-student men who have sex with men—North Carolina, 2003. *Morbidity and Mortality Weekly Report* 53(32): 731–4.

Christensen, H. T., and L. B. Johnson. 1978. Premarital coitus and the Southern Black: A comparative view. *Journal of Marriage and Family* 40: 421–732.

Cooper, M. L. 2002. Alcohol and risky sexual behavior among college students and youth: Evaluating the evidence. *Journal of Studies on Alcohol*, Suppl. 14: 101–7.

Daniluk, J. C. 1993. The meaning and experience of female sexuality: A phenomenological analysis. *Psychology of Women Quarterly* 17: 53–69.

Davidson, J. K. Sr. 1993. Premarital sexual intercourse and axiomatic theory construction. *Sociological Inquiry* 63: 84–100.

Davidson, J. K. Sr., and N. B. Moore. 2004. Unpublished data.

Davidson, J. K. Sr., N. B. Moore, and K. M. Ullstrup. 2004. Religiosity and sexual responsibility: Relationships of choice. *American Journal of Health Behavior* 28: 335–46.

DeLamater, J. D. 1987. A sociological approach. In *Theories of sexuality*, ed. J. H. Geer and W. T. O'Donohue, 237–56. New York: Plenum.

DeRachewitz, B. 1964. *Black Eros: Sexual customs of Africa from prehistory to the present day.* New York: Lyle Stuart.

DiIorio, C., W. Dudley, and J. Soet. 1998. Predictors of HIV risk among college students: A CHAID analysis. *Journal of Applied Biobehavioral Research* 3: 119–34.

Duncan, C., D. Miller, E. Borskey, B. Fomby, P. Dawson, and L. Davis. 2002. Barriers to safer sex practices among African-American college students. *Journal of the National Medical Association* 94: 944–51.

Fennell, R. 1997. Health behaviors of students attending historically Black colleges and universities: Results from the National College Health Risk Behavior Survey. *Journal of American College Health* 46: 109–17.

Ford, D. S., and C. R. Goode. 1994. African American college students' health behaviors and perceptions of related health issues. *Journal of American College Health* 42: 206–10.

Harvey, S. M., L. J. Beckman, and C. Wright. 1996. Perceptions and use of the male condom among African-American university students. *International Quarterly of Community Health Education* 16: 139–53.

Hyde, J. S. 2005. The gender similarities hypothesis. *American Psychologist* 60: 531–92.

Jacobus, S. 1937. *Untrodden fields of anthropology.* New York: Falstaff.

Johnson, E. H., L. A. Jackson, Y. Hinkle, D. Gilbert, T. Hoopwood, C. M. Lollis, , C. Willis, and L. Gant. 1994. What is the significance of Black-White differences in risky sexual behavior. *Journal of the National Medical Association* 86: 745–59.

Johnson, L. B., and R. Staples. 2005. *Black families at the crossroads: Challenges and prospects.* San Francisco, CA: Jossey-Bass.

Kenya, S., M. Brodsky, W. Divale, J. P. Allegrante, and R. E. Fullilove. 2003. Effects of immigration on selected health risk behaviors of Black college students. *Journal of American College Health* 52: 113–20.

Langer, L. M., G. J. Warheit, and L. P. McDonald. 2001. Correlates and predictors of risky sexual practices among a multi-racial/ethnic sample of university students. *Social Behavior and Personality* 29: 133–44.

Laumann, E. O., J. H. Gagnon, R. T. Michael, and S. Michaels. 1994. *Social organization of sexuality: Sexual practices in the United States.* Chicago: University of Chicago Press.

Lewis, L. M., R. S. Melton, P. A. Succop, and S. L. Rosenthal. 2000. Factors influencing condom use and STD acquisition among African-American college women. *Journal of American College Health* 49: 19–23.

Mosher, W. D., A. Chandra, and J. Jones. 2005. Sexual behavior and selected health measures: Men and women 15–44 years of age, United States, 2002. *Advance Data from Vital and Health Statistics, No. 362.* Hyattsville, MD: National Center for Health Statistics.

National Center for Education Statistics. 2005. *Contexts of post secondary education: Minority student enrollments* (Indicator 31). Washington, DC: U.S. Department of Education.

Olson, T. D. 2005. Sexuality education: Philosophies and practices in search of a meaningful difference. In *Speaking of sexuality: Interdisciplinary readings*, 2nd ed. Ed. J. K. Davidson Sr., and N. B. Moore, 375–86. Los Angeles: Roxbury.

Price, J. H., and P.A. Miller. 1984. Sexual fantasies of Black and White college students. *Psychological Reports* 54: 1007–14.

Ragon, B. M., M. J. Kittleson, and R. W. St. Pierre. 1995. The effect of a single affective HIV, AIDS educational program on college students' knowledge and attitudes. *AIDS Education and Prevention* 7: 221–31.

Reinsch, J. M., S. A. Saunders, C. A. Hill, and M. Ziemba-Davis. 1992. High-risk sexual behavior among heterosexual undergraduates at a Midwestern university. *Family Planning Perspectives* 24: 116–45.

Robinson, B. E., K. Scheltema, and T. Cherry. 2005. Risky sexual behavior in low-income African-American women: The impact of sexual variables. *Journal of Sex Research* 42: 224–37.

Robinson, I. E., and D. Jedlicka. 1982. Change in sexual attitudes and behavior of college students from 1965 to 1980: A research note. *Journal of Marriage and Family* 44: 237–40.

Robinson, W. L., and K. S. Calhoun. 1983. Sexual fantasies, attitudes and behavior as a function of race, gender and religiosity. *Imagination, Cognition and Personality* 2: 281–90.

Samuels, H. P. 1997. African-Americans. In *The international encyclopedia of sexuality: Vol. 3. Spain to the United States*, ed. R. T. Francoeur, 1416–19. New York: Continuum.

Sandler, A. D., T. E. Watson, and M. D. Levine. 1992. A study of the cognitive aspects of sexual decision making in adolescent females. *Developmental and Behavioral Pediatrics* 13: 202–7.

Soet, J., W. Dudley, and C. Dilorio. 1999. The effects of ethnicity and perceived power on women's sexual behavior. *Psychology of Women Quarterly* 23: 707–23.

Sprecher, S., and E. Hatfiefld. 1996. Premarital sexual standards among U.S. college students: Comparison with Russian and Japanese students. *Archives of Sexual Behavior* 25: 261–88.

Taylor, S. E., C. Dilorio, T. T. Stephens, and J. E. Soet. 1997. A comparison of AIDS-related sexual risk behaviors among African-American college students. *Journal of the National Medical Association* 89: 397–403.

Upchurch, D. M., L. Levy-Storms, C. A. Sucoff, and C. S. Aneshensel. 1998. Gender and ethnic differences in timing of first sexual intercourse. *Family Planning Perspectives* 30: 121–27.

Valentine, P. A. 2003. Patterns of safer sex practices among allied health students at historically Black colleges and universities. *Journal of Allied Health* 32: 173–8.

Weinberg, M., and C. Williams. 1988. Black sexuality: A test of two theories. *Journal of Sex Research* 25: 197–218.

Whaley, A. L., and E. B. Winfield. 2003. Correlates of African-American college students' condom use to prevent pregnancy, STDs, or both outcomes. *Journal of the National Medical Association* 95: 702–9.

Willie, C. V. 1991. *A new look at Black families*, 4th ed. Dix Hills, NJ: General Hall.

Zaleski, E. H., and K. M. Schiaffino. 2000. Religiosity and sexual risk-taking behavior during transition to college. *Journal of Adolescence* 23: 223–7.

NOTES ON CONTRIBUTORS

J. Kenneth Davidson Sr. is professor emeritus of sociology and past director of family studies at the University of Wisconsin-Eau Claire. He received his PhD in family sociology from the University of Florida and is a fellow in the National Council on Family Relations. Dr. Davidson is also a past recipient of the Ernest G. Osborne Excellence in Teaching Award. His professional publications and research interests have focused on sexual attitudes and behavior of college students, the female sexual response, sexual adjustment and satisfaction, and adoption attitudes of unmarried pregnant teens. The Davidson-Moore Archives at the Kinsey Institute for Research in Sex, Gender, and Reproduction at Indiana University reflect the past 25 years of their collaborative research and publications.

Robert Davis is professor of sociology and chair of the Department of Sociology and Social Work at North Carolina Agricultural and Technical State University, past director of the Institutional Assessment Office, and former vice chancellor for Academic Affairs. He received his PhD in sociology at Washington State University and did postdoctoral work at the University of Wisconsin–Madison in the Institute for Research on Poverty and the Center for Demography and Ecology. Dr. Davis's special areas of interest are sexual attitudes and behavior of college students, suicide, homicide, poverty, and education-related issues, especially at historically Black colleges.

Noah D. Drezner is an assistant professor in higher education at the University of Maryland, College Park. Drezner's research interests include philanthropy and fundraising as it pertains to higher education. Additionally, Dr. Drezner's research focuses on the ways in which minority and special serving institutions contribute to the nation. He has published numerous articles and given several presentations on related topics.

Pamela Felder Thompson received her PhD in higher education from the University of Pennsylvania. She received her MEd from Temple University in educational leadership and policy studies and her BA in English from the University of Maryland, Eastern Shore. Before coming to Columbia University's Teachers College, Dr. Thompson was a lecturer in the Policy, Management and Evaluation Division in the Graduate School of Education at the University of Pennsylvania. She has more than ten years of experience in higher education both in administration and in research. Her fields of interest include graduate education with specializations in diversity, student retention and persistence, and doctoral student development with an emphasis on the African American doctoral student experience.

Jacqueline Fleming is currently director of the General University Academic Center at Texas Southern. At Texas Southern, she has also served as an associate faculty member in the Department of Counseling and Educational Leadership as well as a learning specialist and retention coordinator. She holds a bachelor's degree in psychology from Barnard College and a PhD in personality and development from Harvard University.

Valera T. Francis received her PhD in higher education administration from the University of New Orleans, where her research interests focused on the history of HBCUs in the United States. She began her professional career in higher education at Xavier University of Louisiana, a small, private historically Black university. From Xavier, she moved to the University of Massachusetts Boston, where she was director of the Office of Research and Sponsored Programs. Dr. Francis is now director of the Office of Sponsored Programs at the University of North Carolina at Greensboro.

Shannon Gary is the assistant dean of the Pennoni Honors College at Drexel University in Philadelphia, Pennsylvania. He is currently pursuing his doctoral degree in higher education management at the University of Pennsylvania's Graduate School of Education.

Marybeth Gasman is an associate professor of higher education at the University of Pennsylvania. She is a historian who focuses on African American higher education, specifically Black colleges, fund-raising, philanthropy, and leadership. Her most recent books are *Envisioning Black Colleges: A History of the United Negro College Fund* (Johns Hopkins University Press, 2007) and *Understanding Minority Serving Institutions* (SUNY, 2008). In 2006, she received the Associate for the Study of Higher Education's Promising Scholar/Early Career Achievement Award.

Mark S. Giles is an assistant professor in the Department of Educational Leadership at Miami University, Ohio. Dr. Giles situates his research in the social history of Blacks in higher education. His interests include critical race theory, social-change leadership models, and Black education (P-16).

Gaetane Jean-Marie is an associate professor in the Department of Educational Leadership and Policy Studies at the University of Oklahoma. Her research focuses on Black female leadership in both the higher education and K-12 environments. She has published many peer-reviewed articles in journals such as *Educational Foundations, Journal of School Leadership, Journal of Educational Administration,* and *Women in Higher Education.*

Nelwyn B. Moore is professor emerita of family and consumer sciences at Texas State University–San Marcos. She received her PhD in child development and family relations from the University of Texas, Austin, where she also completed a postdoctoral program in family therapy. A certified marriage and family therapist (American Association of Marriage and Family Therapists), Moore is a past recipient of the Ernest G. Osborne Excellence in Teaching Award from the National Council on Family Relations. Her professional publications and research interests have focused on family life education, sexual attitudes and behavior of college students, adoption attitudes of unmarried pregnant teens, and teen pregnancy/parenting programs.

Kristen L. Safier is an associate in the Litigation Department of Taft Stettinius & Hollister LLP. She received her undergraduate degree from Miami University and law degree from the University of Cincinnati, graduating Order of the Coif. Ms. Safier is also a PhD candidate in educational policy at the University of Pennsylvania. She has taught courses in gender and law at Miami University and legal research and writing at the University of Dayton School of Law. She is a member of the firm's Professional Women's Resource Group (PWRG).

Adam Parrott-Sheffer is currently the director of Instruction for East Chicago Lighthouse Charter School in Gary, Indiana. He has an MSEd in education policy from the University of Pennsylvania and a BA in humanities from the University of Illinois, Urbana-Champaign.

Bianca Taylor holds a BA from the University of Maryland and an MSEd from the University of Pennsylvania. For the past four years, she has worked in higher education and currently works in law school admissions in Manhattan.

Albert Tezeno is currently the executive director of Student Financial Services at Texas Southern University and was formerly the director of Student Financial Services. He holds a BS in computer and information systems as well as an MBA. He is currently pursuing a PhD as well.

Christopher L. Tudico is a PhD candidate at the University of Pennsylvania's Graduate School of Education. He is writing his dissertation on the history of Mexican American higher education.

Amy E. Wells is an associate professor of higher education in the Department of Leadership and Counselor Education at the University of Mississippi and a 2006 Faculty Research Fellow. Dr. Wells's research interests include the history of higher education, and she is coauthoring a book on the history of research universities in the South from 1890–1990 with Professor John Thelin of the University of Kentucky. Her research has been supported by the Rockefeller Archive Center in Tarrytown, New York, and her work about Southern education has appeared in *History of Higher Education Annual, Community College Journal of Research and Practice, Urban Education*, and in *Women and Philanthropy in Education*, a volume edited by Andrea Walton.

Patricia C. Williams is the mother of five and grandmother of six. She began her career in higher education administration at her alma mater, Fordham University, and is now in her seventh year as dean of the W. E. B. Du Bois College House at the University of Pennsylvania. Williams is currently pursuing her doctoral degree in higher education management at the University of Pennsylvania.

Meghan Wilson is a fourth-year EdD student in higher education at the University of Pennsylvania. Her research interests include access to and success in higher education, dual enrollment programs for nontraditional student populations, and the transition from high school to college. She currently is a part-time faculty member at La Salle University in Philadelphia.

Sylvia Zamora is the former coordinator for recruitment and retention in the General University Academic Center, where she oversaw 14 academic support programs and 6 programs for Hispanic students. She holds a BS in biology and an MA in counseling from University of Texas, Pan American. She is also president of the ALTRUSA organization in Houston.

INDEX